Piotr Lipiński and Konrad Świrski (Eds.)

Towards Modern Collaborative Knowledge Sharing Systems

Studies in Computational Intelligence, Volume 401

Further volumes of this series can be found on our homepage: springer.com

Vol. 379. Ferrante Neri, Carlos Cotta, and Pablo Moscato (Eds.)
Handbook of Memetic Algorithms, 2011
ISBN 978-3-642-23246-6

Vol. 380. Anthony Brabazon, Michael O'Neill, and Dietmar Maringer (Eds.)
Natural Computing in Computational Finance, 2011
ISBN 978-3-642-23335-7

Vol. 381. Radosław Katarzyniak, Tzu-Fu Chiu, Chao-Fu Hong, and Ngoc Thanh Nguyen (Eds.)
Semantic Methods for Knowledge Management and Communication, 2011
ISBN 978-3-642-23417-0

Vol. 382. F.M.T. Brazier, Kees Nieuwenhuis, Gregor Pavlin, Martijn Warnier, and Costin Badica (Eds.)
Intelligent Distributed Computing V, 2011
ISBN 978-3-642-24012-6

Vol. 383. Takayuki Ito, Minjie Zhang, Valentin Robu, Shaheen Fatima, and Tokuro Matsuo (Eds.)
New Trends in Agent-Based Complex Automated Negotiations, 2012
ISBN 978-3-642-24695-1

Vol. 384. Daphna Weinshall, Jörn Anemüller, and Luc van Gool (Eds.)
Detection and Identification of Rare Audiovisual Cues, 2012
ISBN 978-3-642-24033-1

Vol. 385. Alex Graves
Supervised Sequence Labelling with Recurrent Neural Networks, 2012
ISBN 978-3-642-24796-5

Vol. 386. Marek R. Ogiela and Lakhmi C. Jain (Eds.)
Computational Intelligence Paradigms in Advanced Pattern Classification, 2012
ISBN 978-3-642-24048-5

Vol. 387. David Alejandro Pelta, Natalio Krasnogor, Dan Dumitrescu, Camelia Chira, and Rodica Lung (Eds.)
Nature Inspired Cooperative Strategies for Optimization (NICSO 2011), 2011
ISBN 978-3-642-24093-5

Vol. 388. Tiansi Dong
Recognizing Variable Environments, 2012
ISBN 978-3-642-24057-7

Vol. 389. Patricia Melin
Modular Neural Networks and Type-2 Fuzzy Systems for Pattern Recognition, 2012
ISBN 978-3-642-24138-3

Vol. 390. Robert Bembenik, Lukasz Skonieczny, Henryk Rybiński, and Marek Niezgódka (Eds.)
Intelligent Tools for Building a Scientific Information Platform, 2012
ISBN 978-3-642-24808-5

Vol. 391. Herwig Unger, Kyandoghere Kyamaky, and Janusz Kacprzyk (Eds.)
Autonomous Systems: Developments and Trends, 2012
ISBN 978-3-642-24805-4

Vol. 392. Narendra Chauhan, Machavaram Kartikeyan, and Ankush Mittal
Soft Computing Methods for Microwave and Millimeter-Wave Design Problems, 2012
ISBN 978-3-642-25562-5

Vol. 393. Hung T. Nguyen, Vladik Kreinovich, Berlin Wu, and Gang Xiang
Computing Statistics under Interval and Fuzzy Uncertainty, 2012
ISBN 978-3-642-24904-4

Vol. 394. David A. Elizondo, Agusti Solanas, and Antoni Martínez-Ballesté (Eds.)
Computational Intelligence for Privacy and Security, 2012
ISBN 978-3-642-25236-5

Vol. 395. Srikanta Patnaik and Yeon-Mo Yang (Eds.)
Soft Computing Techniques in Vision Science, 2012
ISBN 978-3-642-25506-9

Vol. 396. Marielba Zacarias and José Valente de Oliveira (Eds.)
Human-Computer Interaction: The Agency Perspective, 2012
ISBN 978-3-642-25690-5

Vol. 397. Elena Nikolaevskaya, Alexandr Khimich, and Tamara Chistyakova
Programming with Multiple Precision, 2012
ISBN 978-3-642-25672-1

Vol. 398. Fabrice Guillet, Gilbert Ritschard, and Djamel Abdelkader Zighed (Eds.)
Advances in Knowledge Discovery and Management, 2012
ISBN 978-3-642-25837-4

Vol. 399. Kurosh Madani, António Dourado Correia, Agostinho Rosa, and Joaquim Filipe (Eds.)
Computational Intelligence, 2012
ISBN 978-3-642-27533-3

Vol. 400. Akira Hirose
Complex-Valued Neural Networks, 2012
ISBN 978-3-642-27631-6

Vol. 401. Piotr Lipiński and Konrad Świrski (Eds.)
Towards Modern Collaborative Knowledge Sharing Systems, 2012
ISBN 978-3-642-27445-9

Piotr Lipiński and Konrad Świrski (Eds.)

Towards Modern Collaborative Knowledge Sharing Systems

Springer

Editors
Dr. Piotr Lipiński
Technical University of Lodz
Institute of Information Technology
Lodz
Poland

Dr. Konrad Świrski
Warsaw University of Technology
Institute of Heat Engineering
Warsaw
Poland

ISSN 1860-949X e-ISSN 1860-9503
ISBN 978-3-642-27445-9 e-ISBN 978-3-642-27446-6
DOI 10.1007/978-3-642-27446-6
Springer Heidelberg New York Dordrecht London

Library of Congress Control Number: 2011946097

Printed on acid-free paper

Springer is part of Springer Science+Business Media (www.springer.com)

INNOWACYJNA GOSPODARKA NARODOWA STRATEGIA SPÓJNOŚCI
CEWI Platforma informatyczna
UNIA EUROPEJSKA EUROPEJSKI FUNDUSZ ROZWOJU REGIONALNEGO

Preface

Providing easy access to high quality information is essential for academic community. Contemporary system of journal and conference publications appears to be outdated, especially in such domains as computer science. This is mainly due to the fact that process of publishing an article takes several months at least. This is unacceptable in the fields where new technologies arise from one day to the other. Therefore there is a strong need of developing new, fast, more flexible methods of sharing knowledge between scientists and industry.

In this book a new approach of sharing knowledge is proposed. The main idea behind this new approach is to take advantage of collaboration techniques used in industry to share the knowledge and build teams which work at the same subject at different locations (different universities). Large companies are able to create and manage global teams of hundreds employees which are located in different countries. We suggest using the same approach and tools for creating and managing research communities to join global forces of scientist sharing the same interests. This will allow to accelerate the exchange of information between scientists and allow to build global teams of researchers who deal with the same scientific subjects. Furthermore, an easy access to structured knowledge will facilitate cross domain cooperation.

The first part of this book introduces the concept of a platform which can be used for scientific cooperation. It also familiarizes readers with new concepts and technologies which are used in the platform. In the first chapter the authors describe the new approach of sharing knowledge in scientific community, give technical details of the system and technologies which are used. In the second chapter authors present the foundations of managing CAD objects in the engineering environment with PLM systems usage. This chapter is mainly dedicated to those platform users who take advantage of CAD software. Once the main functionality of the platform has been described the access control management using extended RBAC model is presented. The author presents a software tool for managing the security of enterprise information system on the access control level from the point of view of security administrator. The security administrator is responsible for the management of access control of users to data stored in the system. In the following chapter practical information about preparing documentation in DocBook 5.0 is given. The study may be useful for people who need to choose a format and tools for creating and organizing scientific/technical documents or teaching materials. In the second part of the book examples of the projects which are developed using the platform are given. The first three projects focus on computer graphics. The first one describes augmented reality technology which is used for passing the information about works of art and architectural monuments. Continuing with computer graphics, in chapter six 3D scene exploration with external controllers is presented. It demonstrates selected interaction methods and their implementation. Next chapter introduces implementation of new photon mapping algorithms. It presents several algorithms for high-performance rendering engines. Authors describe the development of global illumination methods with particular emphasis on photon mapping and consider the possibility of further improvement of these methods. The following four projects described in this book are related to signal processing. In chapter eight the authors propose a do-main-specific language, called TDL, for describing systems clearly, compactly, and quickly. In this chapter authors present an associated compiler, which converts TDL descriptions into Java/C++ implementation code, so as to save programming effort. In particular, the dataflow and data structures are discussed. In chapter nine system design for biomedical signal and vibration analysis and monitoring of machines, based on the Mallat and Hwang wavelet singularity analysis is introduced. Next chapter is devoted to the problem of designing an optimal systolic array for morphological processing. The paper gives the proposal of the new systolic architecture with the examples of using new features. In the following chapter the author outlines the method and application for adaptive wavelet synthesis. This application should be of particular interest for those scientists who use wavelet analysis.

The last part of the book covers three computer networks projects. In chapter twelve the author presents credit-based flow control solution for IP networks. In contrast to well known solutions that utilize flow-level control, the solution is designed as link-level control system, where every set of directly connected network nodes creates a separate control system. Next chapter presents a systematic, computer aided method supporting network planning for interior wireless networks based on IEEE 802.11 standards family. Detailed specification of features and functionalities of the proposed network planning tool available as a web service is provided together with directions of its possible expansion. The book ends with a chapter describing the concept of the universal snapshot file which can be used in distributed environments.

We expect this book will be of interest for researchers, practitioners and industry employees in computer science and mechanics. The editors would like to thank the contributors of the book for their effort in preparing the chapters. We would like to give special thanks to Editor of the Series prof. Janusz Kacprzyk for his help and his support. Thanks to all the people who were involved in completing the book . We hope that our work will contribute to the evolution of new technologies of scientific cooperation.

Piotr Lipiński

Contents

Information Platform TEWI

Piotr Lipiński and Konrad Świrski

Summary

Information Platform TEWI (Technology, Education and Research, Knowledge, Innovation) is an interregional research network, which aims at integrating Polish scientific community using Product Lifecycle Management software. It is being created by three universities: Technical University of Lodz, Warsaw University of Technology and Bialystok University of Technology. The platform TEWI will allow to take advantage of new information technologies and state of the art solutions in research and industry. It will facilitate cooperation between the universities and industry by utilizing compatible software. The Product Lifecycle Management software will make knowledge management easier and more available for scientists. Moreover, the new software will improve the skills of scientific community and make the technology transfer to the industry more fluent. As a result this should make the Polish industry more competitive.

TEWI - four main lines of action:

- (T) Technology - one of the most modern technology in the world of computer design and product lifecycle management as a platform for design and product management that can be used in various sectors of industry (automotive, aerospace, electronics, industrial, pharmaceutical and research);
- (E) Education and Research - comprehensive use of software and systems in the process of research and training can modernize education, for example 3D design tools including computer simulation systems of machine parts and equipments;
- (W) Knowledge - investment in knowledge and human capital, raising the professional qualifications, promotion of the latest technological solutions and creating value-added through team work, computer aided design process, management of versions and modifications, better information flow;

Piotr Lipiński
Institute of Information Technology, Technical University of Lodz
Wolczanska 215, 90-924 Lodz, Poland
e-mail: piotr.lipinski@p.lodz.pl

Konrad Świrski
Warsaw University of Technology, Institute of Heat Engineering
Nowowiejska 21/25, 00-665 Warsaw, Poland
e-mail: konrad.swirski@itc.pw.edu.pl

P. Lipiński, K. Świrski (Eds.): Towards Modern Collaborative Knowl. Sharing Syst., SCI 401, pp. 1–10.
springerlink.com

- (I) Innovation - creating conditions to pursue advanced work of Research + Development, creating innovative products, cooperation between science and industry, stimulating development in the region. The platform will allow to design, create, modify and improve products.

1 Introduction

The contemporary development of science and technology leads to arising more modern and complex systems. Several years ago it was possible to create systems that could be easily controlled or designed by a single person. Today, the complexity of even a simple system is very high, as a result it is difficult for a single per-son to design or execute it. Therefore, it becomes increasingly important to enable cooperation between people in order to create complex systems.

In the recent times the product management process has be-come more important, because of the significant costs of introducing new products due to high expenditures for the project, implementation, advertising, etc.. Nowadays, it is not only necessary to design the system properly, but also to implement it, maintain, develop and withdrawal from the market. The whole process of managing a product is called "product life cycle" . Every modern, complex sys-tem that is working reliably follows certain life cycle. There are specialized software tools, called Product Lifecycle Management (PLM) software which helps managing product life cycle. It includes tools for:

- collaboration between individuals making up a system (at the stage of its creation)
- coordination
- implementation,
- operation
- withdrawal from the market.

PLM software is successfully used in industry, but it is not used at universities, neither in teaching nor in scientific research. The survey conducted at the Technical Universities of Lodz, Warsaw and Bialystok suggest that the scientific world is developing in a less structured way than the industry and it is focused more on the individual, with a significant loss for it. In our opinion, the use of PLM to support research workflow by enabling the exchange of experience and solutions between scientific bodies in Poland, regulation of scientific work by creating documentation and using the same tools that are used in industry will accelerate the development of science in our country.

The current system, which is based on printed publications and conferences, seems to be outdated, it does not sufficiently integrate the scientific community and does not provide adequate diffusion of scientific results to industry. For this reason we have organized a consortium which has set a goal to build an interregional PLM platform, which aims at transferring knowledge between universities and industry. This will lead to integration of the scientific community in Poland through the use of technologies that are used in industry for product lifecycle management. Our platform will enable the collaboration between scientists working in different regions of the country thanks to the centralized structure of the platform and repositories of data and documents with remote access. It will also improve cooperation between scientists working in different fields by using the same platform for both areas: IT and mechanics.

2 Information Platform TEWI

The Information Platform TEWI project is a direct technological solution (computer network and software) that enables designing, building, managing products life. TEWI platform makes it possible to use the infrastructure in the research process in the most technologically advanced sectors of the economy.

The platform is interregional, but at first the main components will be located in Lodz, Warsaw and Bialystok. In each of these locations the basic research and development infrastructure will be placed in computer centers of project's participants. When new local partners join the project, this infrastructure will be also placed in their region. The core servers of the Platform, including servers with licensed software, will be located in the Computer Center of the Technical University of Lodz. The unlimited number of partners can join the TEWI platform therefore the performance of the equipment in the main location in Lodz must be oversize in the beginning.

3 Hardware Description of the Platform

In the technical sense, Information Platform TEWI is an interregional network with computer software. Basic computer system servers with software are being installed in the Information Centre of the Technical University of Lodz. The access to the applications will be available through the broadband network and the Virtual Private Network (VPN) from each project's partner site, in each region (lodzki, podlaski, mazowiecki). In the subsequent stages of development Platform can extend its range over other academic units.

In terms of investment and infrastructure, Platform TEWI consists of the following components:

- Data Centre of Platform, located in the Computer Centre of the Technical University of Lodz,
- Local University Access Centers of Platform in the Computer Centre of the Warsaw University of Technology and the Bialystok University of Technology
- Local Access Centers of Platform partners
- Specialized laboratories of scientific research

Data center platform has already been purchased and is being installed. It includes:

- dedicated file server: IBM System Storage N6040 Model A20 (P/N: 2858-A20)
- TS3310 Tape Library (P/N: 3576-L5B),

- TS3310 Tape Expansion Module (P/N: 3576-E9U)
- IBM DS4800: DS4000 EXP810 Expansion Unit Model 81 (P/N: 1812-81A) with FC hardisks
- IBM DS4800: DS4000 EXP810 Expansion Unit Model 81 (P/N: 1812-81A), with 16 hard disks 1TB SATA
- DS5020 Midrange Disk (Dual Controllers) P/N: 1814-20A,
- EXP520 Expansion Unit P/N:1814-52A
- Firewall
- Set of CTC Union converters
- 4 Blade servers: IBM HS22

TEWI Platform is scalable and modular and must be easily customized, because we assume that it will grow in the future as more universities will participate in the project. The architecture of the system facilitate the system administration from the perspective of IT infrastructure. The access of end-users (graduate students, researchers) to the resources of the platform (software) will be possible from any location through the appropriate access rights.

4 Software

In the software layer, the specialized software of TEWI network will consist of the following components:

a) Integrated management of product lifecycle software
b) Product visualization software
c) 3D CAD/CAM software
d) Engineering calculations software
e) Automatic creation of documentation software.

4.1 Directing PLM Software (Product Lifecycle Management)

PLM software is a modern, advanced, expandable and scalable tool which allows to control the product lifecycle in an integrated way, from the design phase to its natural withdrawal from the market. Using the "thin client" technology, as well as an easy and user-friendly interface, using a web browser WWW it has access to all features available in the software.

PLM software is very important in the process of increasing the competitiveness of companies through the continuous improvement, automation of business processes and procedures, as well as digital modeling. It provides a systematic approach to the configuration, management, the re-use of products structure and it also combines them with appropriate content, such as CAD files, documentation, calculations and illustrations.

All products of PLM software are integrated; they share a common database structure, business objects, process models and user interface, as well as a single login to the system. Completely web software architecture allows communication, both members from different departments within one company and the external users, such as suppliers, partners and customers. Architecture of PLM software allows also smooth data exchange with other applications and systems, such as ERP (Enterprise Resource Planning), SCM (Supply Chain Management) and CRM (Customer Relationship Management).

As the most comprehensive solution in the industry PLM software can provide functionality which suits business needs. If basic ability of product data management, optimization process to industrial standards or support for global product development is needed, PLM software will help to achieve all the requirements. PLM software is used by companies that want to manage products as efficiently as possible throughout their lifecycle.

PLM software is a preconfigured system, which means that after the installation it is ready to work. Preconfigured environment includes the best proven practices: workflow processes, roles and groups of users, reports, data sets, properties of documents and other components, forms and lists of values.

Configuration of the above resources is optimized to perform specific tasks such as managing, reviewing, approving, publishing and processes of making changes in the documentation. All system resources can be freely configured by modifying the existing settings or introducing new settings. Modification of PLM software environment is done by wizards.

Benefits of using PLM software:

- Assures streamlines the process of information flow within the company, provides centralized access to information and assists with storage of information.
- Uses the Digital Product Value Chain to create a chain of connections leading to the best collaboration between all groups.
- Introduces a virtual version of the real product (Digital Product) that gives us some freedom to create by marginalized costs associated with the introduction of products.
- Provides a centralized source of information about the product and its contents; increases the efficiency of the product development; reduces number of errors, corrections and reduces time.
- Gives the possibility of a complete product definition and cooperation helps to understand information within the company regardless of its source.
- accelerates the introduction of the product to the market and reduces the cost of the project.
- enforces compliance with standards
- provides a solid, high-performance technology.

4.2 Multi-CAD

The PLM software has the functionality of multi-CAD, which allows you to manage data from various backgrounds. It includes all the leading CAD applications such as: Solid Edge, Unigraphics NX, Catia, Pro/Engineer, SolidWorks, Inventor,

AutoCAD. Data from different CAD applications can be integrated and managed in a single product structure.

4.3 CAD Software

CAD software is a fully parametric, integrated CAD / CAM / CAE / PDM / PLM system. This software is one of the leading products for 3D design, promoting good habits in the design process and ensures compliance with industry and company standards. CAD software is an easy to use tool, available in the form of different packages; it facilitates adaptation to the specific needs of the company:

- CAD Foundation XE software – cost-effective 3D CAD system contains the basic design capabilities.
- CAD Advanced XE software – CAD 3D, data management and functionality extended by elements such as: advanced editing or wiring design.
- CAD Enterprise XE software – Product Development Sys-tem that allows seamless integration between the team and extended supply chain.

Benefits:

- Unsurpassed geometry creation capabilities allow greater individualization and better product formation.
- Complete integration of the process of application development
- Automatic propagation of changes in the project to all de-pendent components eliminates problems with the translation of data
- The function of the virtual simulation can improve the product performance and can help to achieve the product quality requirements.
- Maximum production capacity is possible due to the automatic generation of working draft, installation instructions and machine code.

CAD Mechanica

This option allows performing strength and thermal analysis based on models created in CAD program.

Application:

- Creating solid, coating, beam or mixed model.
- External Loads: forces and torques, pressure, mass forces, gravitational force, reaction forces, temperature.
- Defining the boundary conditions - bonds: blocking movements, using symmetry model etc.
- Defining the initial conditions.
- Possible standard analysis:
  - Static (in the linear range of the material stiffness for a permanent Yung module) – finds tension distributions, strain, movements, reduced tensions, strain energy and more.

- Contact
- Modal - finds frequencies and forms of own vibrations,
- Buckling - finds critical load of compressed slender rods, walls or coatings,
- Thermal – steady-state – temperature distribution, heat flows, etc.
- Sensitivity – it can be used to study the influence of different model parameters (dimensions) on the resulting parameter analysis,
- Optimization – is used to search for optimal solutions in the domain of selected model parameters and minimize or maximize the objective function (weight) expressed by the resulting parameter analysis.

4.4 Software for Engineering Calculations

Engineering calculations software is the most widely used software for performing, documenting and sharing engineering calculations. Unique look of this software and interface simulating notebook combines a standard mathematical notation, text and graphs in one sheet, creating an ideal tool for acquiring knowledge and reusing calculations and engineering collaboration.

Unlike the other commercial tools, software for engineering calculations allows for the simultaneous design and documentation of calculations using the extensive functionality of the mathematical module and dynamic calculations, taking into account the physical units. One of the advantages of the engineering calculations software is user-friendliness. The interface is intuitive, many operations can be performed by using the mouse, equations and algebraic expressions are displayed in the graphic form, rather than in the text form.

Benefits:

- Easy to use – requires no special programming skills
- Increases productivity, saves engineers time and reduces the number of errors
- Improves verification and validation of the critical calculations
- Promotes good habits, facilitates the calculations re-use
- Complete documentation of calculations supports compliance with standards.

4.5 Software for Dynamic Documentation and Publication

Providing an easy access to high quality information is essential to ensure the financial success for the organization and the ability to meet customer requirements. Relying on traditional publishing software to meet the needs of business information can be costly and error-prone. Traditional publishing software contributes to inconsistencies in sequence and structure of information, forcing excessive, unnecessary processes and often requires updating the document manually. Too often the bottleneck of the publishing and development process causes delays in launching product to the market and forces organizations to release the unfinished, inaccurate or outdated information that ultimately results in loss of market opportunities, high maintenance costs and low customer satisfaction.

Software for dynamic documentation and publication increases efficiency and automates your publishing information process and precludes the inefficiencies that can be costly for the company. By putting pressure on the creation of XML documents, technical illustration creation, content management and the process of its creation and dynamic publishing, you can improve your publishing processes and help organization to gain a significant competitive advantage and differentiate themselves from other products of this type.

Benefits:

- Customer satisfaction is achieved by delivering high quality personalized documentation that is logical and accurate.
- Cost cutting through the dynamic publication of your trade information to a variety of formats – it can be printed and delivered in electronic format – from only one source.
- Elimination of excessive work by increasing efficiency of documents creation, design of technical illustration, translations, reviews and production of publications
- Increase of an employee productivity – better organized workflow
- Speeding up the process of compliance by faster creation of more complete documentation, accurate tracking of documents and automated documents workflow
- Delivering products to the market faster by creating documentation parallel to the process of product development.

4.6 Visualization Product Software

Product development requires diverse information from many different sources. One of the most significant issues that many manufacturers must face is how to efficiently share the huge amount of information about the product that is formed during the development process. Visualization product software offers better integration with PLM software, CAD, many applications such as MCAD, ECAD and over 200 other types of documents. With the visualization product software everyone can: see, format, interact and cooperate with all forms of the digital product data.

Opportunities:

- Provides access to information about the product to all team members in web and scalable environment.
- Allows observing data about the product (MCAD/ECAD files, MS Office documents) without the application used to create them.
- Accurately measures, divides and compares 3D models, allows collecting parts and complicated product structures.

- Enables collaboration on a centralized digital information about the product by tags and real-time collaboration.
- Improves performance by enabling users easy sharing information in the visual form.

4.7 Management of Products and Data Sources Software

Creating and efficient products management

Product creation is a complicated and long-term process. We want to be able to easily change its properties and features as well as create a large number of copies of the product. They may differ radically, depending on their destination. This is where the proposed software becomes irreplaceable.

Management of products and data sources software is a part of PLM adapted to the needs of the company. It offers the ability to create and efficient products management, from the design phase.

Opportunities:

- Controls over the development process at every stage
- Creating a single source of information
- Control over approvals and changes
- Collaboration between groups.

4.8 Summary of the Technical Design Assumptions of the Information Platform TEWI

The technical solutions ensure the creation of modern inter-regional network infrastructure; provide access to computer applications that support scientific research. The proposed architecture and investments are characterized by the use of modern technical solutions (network, server architecture, links between scientific institutions) and cost optimization. We utilize existing infrastructure as well as human resources of the Computer Centre of the Technical University of Lodz in managing extensive computer networks. The proposed hardware investments and the supply of the intangible as-sets are justified and allow using long-term investments together with the optimization of investment funds.

- The solution (Project TEWI) is in line with EU horizontal policies.
- The proposed project is a modern use of available information and communication techniques (ICT). The operation of the Information Platform TEWI is based on the concept of ICT and also on introducing the most innovative methods of conducting research using computer methods that are used in the interregional solutions.
- The project is consistent with the recommendations of the EU as well as with the existing national regulations.
- The technical project solutions are consistent with the best known practices and are characterized by the optimal price-quality ratio. In

the preparation of the project some variants (ineffective technologically and in terms of the price) were rejected, for example:

- Distributed architecture of the platform (separate plat-form for each partner).
- Distributed end-user software installation.

5 Conclusions

The authors of the project hope that the platform will enable introduction of new technologies and innovative solutions to the process of research and development. Collaboration between universities and industrial units and will contribute to the transfer of knowledge from universities to industry through the access to the product lifecycle management software. Access to the computerized processes will enable the dissemination of knowledge and information management. Equipment of the research process with he new technology will contribute to more attractive offer and increase the number of researchers and graduate students at technical universities that will promote creation of skilled labor to support new segments of the economy - high technologies, innovations. The impact of research on the growth of economic competitiveness will increase while the supra-regional cooperation will help to enhance competitiveness, knowledge and skills in the regions and in the country.

In the following sections we demonstrate the IT projects which will be first developed at the Technical University of Lodz, Warsaw University of Technology and Bialystok University of Technology. The experience in developing this software using TEWI platform will be used to adapt the TEWI platform to the IT project requirements in scientific environment at a university.

References

[1] Li, H.-Y., Liu, X., Feng, G.-Q., Wang, C.-E.: A Web-based PLM system research and implementation in a colla-borative product development environment. e-Business Engineering. In: IEEE International Conference on ICEBE 2005, Beijing (2005)

[2] Xia, Y., Yang, Z., Zhang, L., Hu, L., Zhang, Y.: The study on common/unique issues in PLM/CAD integration and a design of flexible integration architecture. In: The Fifth International Conference on Computer and Information Technology, CIT 2005 (2005)

[3] Res. Inst. for Rationalization & Oper. Manage., Univ. of Aachen, Aachen.: Next generation PLM-process management for the development of telecommunications products in the multi-lifecycle environment Management of Innovation and Technology. In: 4th IEEE International Conference on ICMIT 2008 (2008)

[4] Thomas, V., Neckel, W., Wagner, S.: Information technology and product lifecycle management. Electronics and the Environment. In: Proceedings of the 1999 IEEE International Symposium on ISEE (1999)

Introduction to Integration of PLM and CAD Systems within TEWI Platform

Bruno Deszczyński

Abstract. This article presents the foundations of managing CAD objects in the engineering environment with PLM systems usage. In order to do the above, model of managing change and configuration within TEWI platform has been described. In addition the paper introduces typical methodologies of designing products with CAD systems.

1 Introduction

The purpose of the TEWI platform project (where TEWI is a polish abbreviation of Technologia, Edukacja, Wiedza, Innowacja – Technology, Education, Knowledge and Innovation) is to establish a scientific network that connects university units and provides the academical society with access to the enterprise information systems. The application of best in class software systems – integrated 3D CAD/CAM with PLM[1] (Product Life Cycle Management) systems and advanced applications for mathematical engineering and automation in CAD projects documentation creation will allow to deliver scientific activity in all university nodes on the highest possible level. The research and development that will be using TEWI platform will surely develop multiple applications for commercial or university use and will help to create innovative products. More details in case of TEWI platform's creation can be found under [4], where the feasibility study describes initial specification of the platform. PLM which is the core of the platform has been initiated by the need for broad-based product management within the manufacturing industry. Previous systems for EDM (Engineering Data Management), and later PDM (Product Data Management) were not able to meet the

Bruno Deszczyński
Warsaw University of Technology, Institute of Heat Engineering
Nowowiejska 21/25, 00-665 Warsaw
e-mail: b.deszcz@itc.pw.edu.pl

[1] PLM – is the process of management whole product lifecycle, from sketch and design, through manufacturing, to product retreat and utilization. It also includes also post-sale analysis, servicing, concept and change management etc.

P. Lipiński, K. Świrski (Eds.): Towards Modern Collaborative Knowl. Sharing Syst., SCI 401, pp. 11–22.
springerlink.com

demands of supervising engineers, CAD (Computer Aided Design). PLM is the process of managing the entire product lifecycle, from concept, through design and manufacturing, to service the system until the withdrawal from the market [3]. PLM integrates people, data, processes and business systems, and forms the basis of information about products within companies. Modern PLM systems also include extendable workflows or program management capabilities, standardized monitoring status of the projects which significantly accelerates product management.

Product Lifecycle Systems are the most comprehensive solution in the industry which offers the functionality of any business' needs. If there is a need for product data management, optimization of industrial processes to the standards and support for global product development, PLM software helps to achieve all the requirements. To achieve the goals mentioned, the Windchill system has been deployed on Technical Universities of Łódź, Warszawa and Białystok in an virtualized infrastructure. Virtualization is an important factor for the infrastructure part of TEWI platform as it enables many obvious advantages like remote access, better testing and development environments, consolidation of servers, dynamic load balancing and disaster recovery.

1.1 Integration of PLM and CAD Systems

Have you ever wondered what the milk containers have in common with mobile phones, rockets or planes? The answer is that all of them have been designed with CAD tools. Additionally, to optimize the production process there was a need to implement the Product Lifecycle Management system (PLM). It's hard to imagine the possibility of effective design of such complex objects as a plane without properly working integrated systems. In case of less complex products, e.g. boxes or household equipment, the essential factor which determines the choice of production support system is plurality of configuration of an object (e.g. changing of parameters from series to series, design variants, etc.). In order to control huge amount of data during the design phase and production, it is necessary to use change control and configuration management system as well as secure sharing of various types of data. A solution which can provide all mentioned higher functionalities is Windchill PDMLink created by Parametric Technology Corporation (PTC).

Software from PLM domain, for instance Windchill, has been designed to enable complex management of sophisticated products' configurations and to provide the platform to control, cooperate and communicate information about the product. The greatest challenge is to control the information need, starting with the requirements gathering phase, through design, right up to production and service delivery. Working with the product requires involvement of many different individuals, varying from each other with the scope of responsibilities, and therefore their needs determined by access privileges and data needed to process. Despite the differences above, whole set of data has to be synchronized to effectively realize the process. For example, engineers have to be provided with access to the latest versions of the designs, while the marketing department works on product visualization, updating advertising materials, manuals etc. Effective control requires well-considered process, which won't overload the product creators with additional work and duties.

2 Continuous Changes and Adaptation of Change and Configuration Management on the Universities

Nowadays Change and Configuration Management is still a huge problem for production companies. Configuration Management is a process of managing decisions connected with the product definition in order to assure it's coherence in the design, through verification, launching to the market, distribution and service. On the other hand Change Management is the process of controlling and tracking the changes within the project and observation of how the product evolves during the whole lifecycle. Both processes are fundamental in order to assure effectiveness of designing and production as well as high quality of produced goods. The negligence of the above may cause some delays with launching the product to the market, over-budgeting and quality problems resulting from de-synchronization of information about changes and lack of possibility to verify decisions in the project coherently.

Effective change and configuration management requires two basic elements:

- close cooperation between working teams and coordination of their activities,
- keeping clarity of the needs in every single phase of lifecycle,

PLM software assures obtaining the single source of information about the product. Thanks to that Technical Universities participating in TEWI platform may practically use the easy way of entering data to the common repository[2] in the form of Windchill, accessible through the WWW interface. Thanks to the solution called Windchill Workgroup Manager, the Windchill software provides support for the most popular CAD programs on the market. The scope of supported applications contains: Pro/Engineer, AutoCAD, Autodesk Inventor, CATIA V5, Solid Works, NX, Mathcad, Arbortext IsoDraw, CoCreate Drafting. Usage of common interface enables unified management of CAD data, independently from the tool used for its creation.

[2] Repository - a software repository is a storage location from which software packages may be retrieved and installed on a computer.

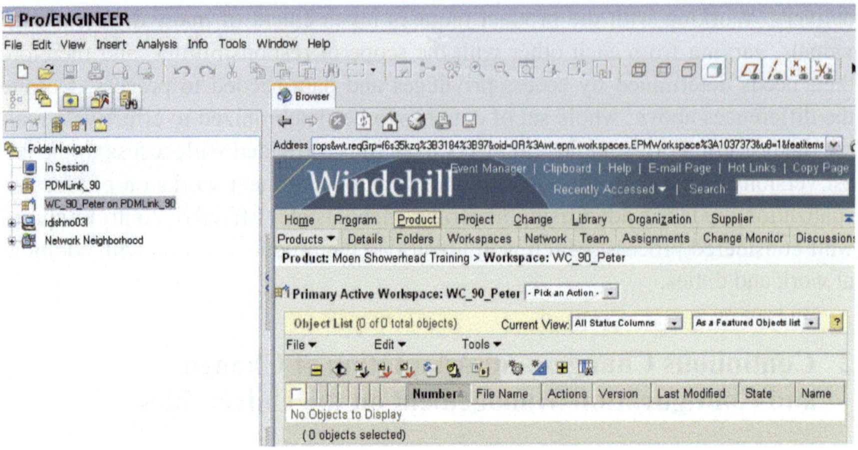

Fig. 1 Pro/Engineer: connecting to Windchill system form CAD tool integrated browser

Presented on the above picture solution integrates the standard set of functionalities available for Windchill users (from the level of a web browser) within the graphic interface of a CAD tool. For instance, the user can open CAD documents directly from the Windchill server, saving data in the common repository, integrating local changes with the modifications made by other users or renaming the documents. That approach to the project data management resolves many logistic issues, which production companies have to struggle with.

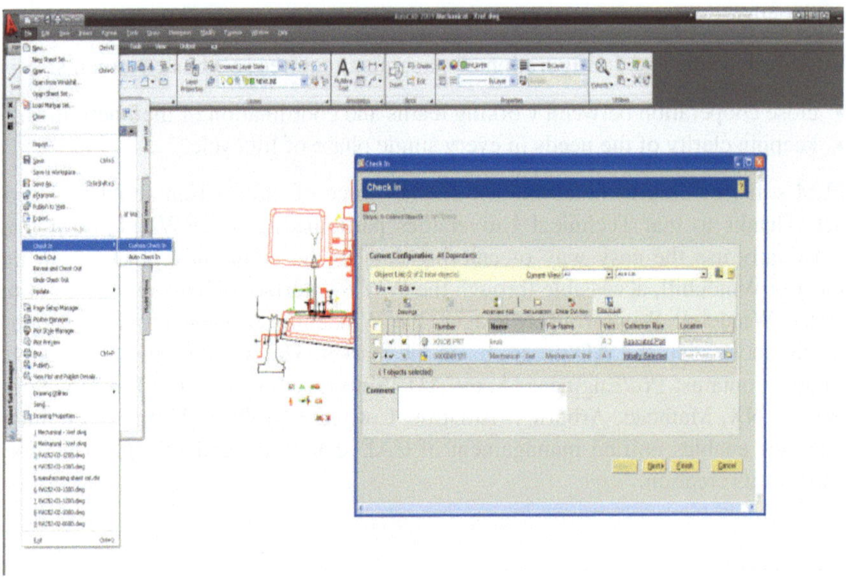

Fig. 2 AutoCAD: entering data to the server

2.1 Configuration and Change Management in TEWI's Windchill

Windchill system contains mechanisms enabling managing changes, covering all concerns connected with introducing formal and informal changes to the data, which are a matter of company activity. For example, the tester may detect malfunction in one of the components, and then the formal process implemented in the system leads to removing identified defect with the parallel verification of that change. Change management process existing in the Windchill system is based on the standard industrial processes, widely described in [1].

It contains few different types of objects representing the change and its consequences:

- report about the problem,
- change request,
- change execution instructions with the implementation tasks – called change notice.

Additionally, all the objects are subordinated to standard processes of acceptance, verification and approval, which allows employees involved in the project to track the changes made on the product in the structured way.

Change management process has been facilitated with the implementation of direct connections between the change notification, components and assemblies, which need those changes. Therefore, there are no issues with identifying the data, which are modified, and additionally engineers involved in the process can analyze the modifications in the CAD data on a current basis (e.g. before acceptance of the activities, which are about to imply changes to the components in the change process)

Configuration management, in case of Windchill system, relies on the possibility of defining a set of parameters – configuration specifications – typical for certain objects, and then using it to filter the displayed data in the dynamic structure of the product. Such specification defines the collection of specific components' versions and assemblies of the structure, which are displayed to the user and/or will be used to navigate over the structure of the data connected with the CAD documents.

Standard configuration specifications, which may be chosen by the engineer reviewing the structure of the product is for example the specification presenting the latest versions of design data. Another way to mange the structure is to use baselines - which can be defined as captured in the certain time snapshot representing the appearance of the product in the past – e.g. before implementing the changes based on the tests verifying engine hp. It is also possible to use a set of parameters modifying the appearance of the product structure, such as "effectivity" (defining when the product was/will be used) of a certain component version. "Effectivity" can be specified based on attributes like date, serial number or LOT number.

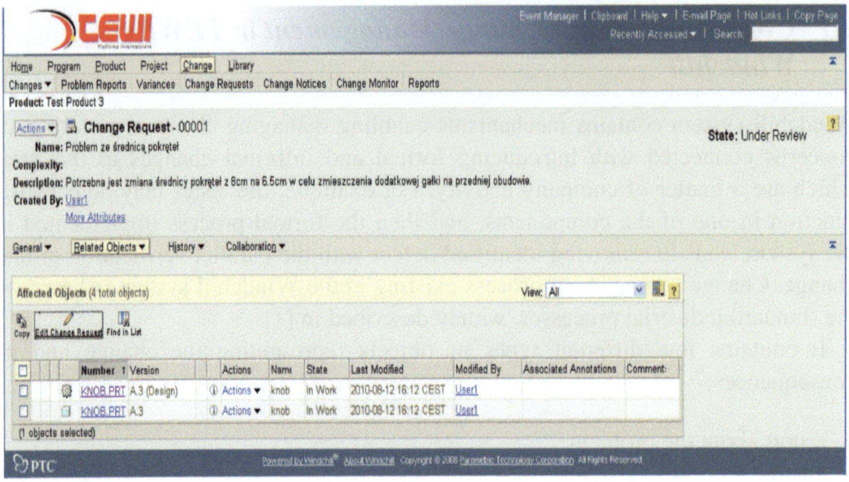

Fig. 3 Affected objects of change request in case of CAD data.

The next factor which may be used to modify the structure of the product is the product lifecycle state – in example presenting only the incomplete components or just released to manufacturing objects. The lifecycle state additionally shows the status of the product and gives the possibility of informing about the current status of the design process. Additionally, the mechanisms of configuration management can be used to define views of the structure – e.g. engineer project view (Design) or view of the structure for the production (Manufacturing).

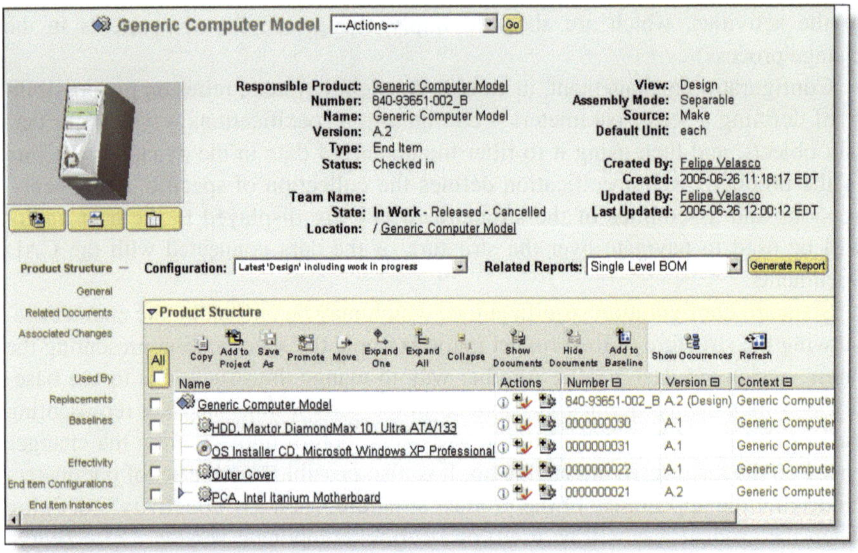

Fig. 4 Product configuration management

Manufacturing view may differ from the design view with the existence of components specific for production purposes – e.g. parts of container, precise number of elements appearing - instead of just 1 version (4 wheels instead of 1). All the structure view variants mentioned above enables access to the specific CAD data, stored in the repository of Windchill system. This assures strict control of the data accessible by the employees of various departments of the company as well as subcontractors – depending on information demands, access rights or tasks allocated. An obvious advantage is also the fact that the different views provide just the needed amount of information – no overloading with information is happening.

2.2 Distinctness of the Characters – Various Tools and Data Formats in Change and Configuration Management

Nowadays not only the geographical diversity of production companies and universities is the problem, but also the number of units involved in the design process and then the production. It is also possible that the subcontractors are restricted with using only a certain specific type of CAD tool which brings additional constraints. The other example of coexistence of various representations of the same object in various design applications is the necessity of defining the interfaces in the interconnection of mechanical and electrical projects – MCAD/ECAD.

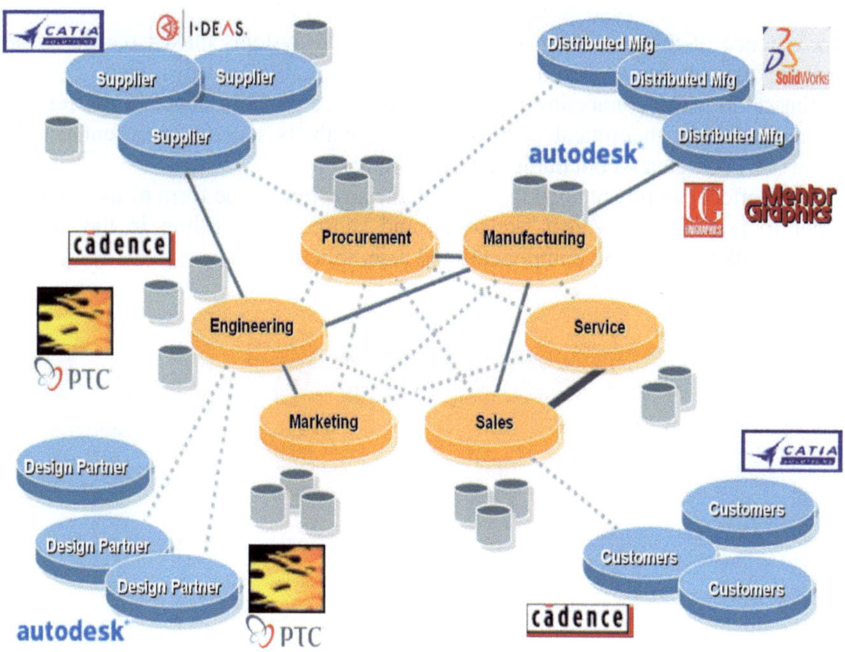

Fig. 5 Babel Tower of CAD domain

There are two main types of designer environments, in which the data of various CAD applications are alternating – the subject is widely described in [2].

First of them, homogeneous, can be characterized with the situation when within one company there is a need to use specific data formats for few products/tools, however there is no direct connection between them among one structure. The situation can be observed when the archive data are being stored because the cost of migration to the new CAD systems strongly exceeds the regular cost of service, maintenance, etc. Another factor determining the necessity of maintaining CAD data in various formats is diversification of the tools in order to the best utility for the certain line of products.

The second type of designer environment are heterogeneous environments – i.e. single components of one product structure are used in various applications or have various representations. One product is not described within one CAD tool and certain components have to be analyzed with the different set of designer tools. In example the producers of interior elements for the automotive sector, who use ready-made components – providers offer visualization of the parts only in a certain format. The other example can be libraries of standard catalogue parts. On the other hand the clients may require providing the designs in a certain format, what implies usage of the certain tools to ensure smooth cooperation.

3 "Top-Down" and "Bottom-Up" Methods of Design

Coexistence of the data from miscellaneous CAD systems makes the designing process difficult, however it does not make it impossible anyways. One of the potential solutions is connecting the general structure of the product in the form of the collection of theoretical elements (parts), with its specific representation by means of CAD data – containing meta-attributes.

Such effect can be obtained by creating the structure in the form of meta-data in order to prepare the project skeletons, and subsequently filling in the general "form" with the concrete components and assemblies.

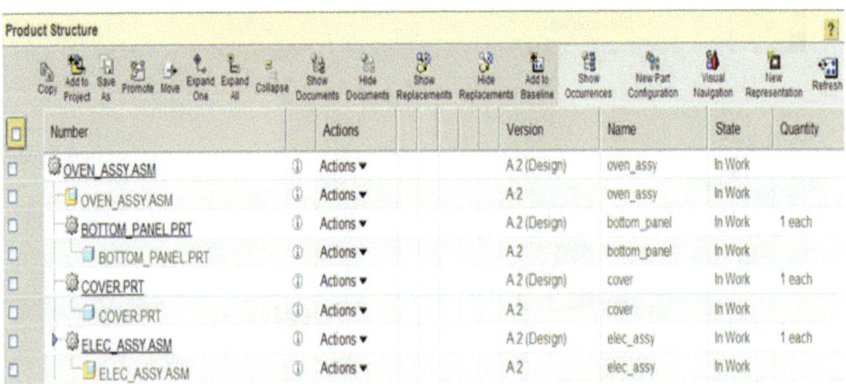

Fig. 6 Structure with CAD documents

This approach is called top-down design. In that case the design process is controlled with the higher level of abstraction – the framework of the product structure is created in the Windchill system, when the designing of the certain elements is delegated to the separate work groups. Such activity conducts to the natural distribution of work and is easier to manage in case of checking the work progress.

There are many possibilities of integration of CAD representation with the previously defined structure in case the data are entered to the system from the CAD tool level (e.g. Pro/Engineer). The connections between the CAD object and the representation may be determined based on a list of attributes like the names, serial number, file name etc., and thanks to that CAD documents will be automatically referenced with the appropriate data of the theoretical framework. The alternate solution is manual connection of meta-data with the CAD documents – i.e. via pointing out the appropriate linkage between theoretical part and corresponding component or compound.

Main advantages of "top-down" method are:

- possibility of early defining the interfaces between components
- partial automation of assemblies
- facilitated control of the changes being in conflict within the project
- well-defined communication of the sketched intention of the whole project

Another approach is the "bottom-up" method of design. It's described with the scenario when the CAD design is already defined (even the initial version or multiple variants of CAD product designs are available), and afterwards the groundwork in the form of meta-data within PLM system is created. In that case the necessity of

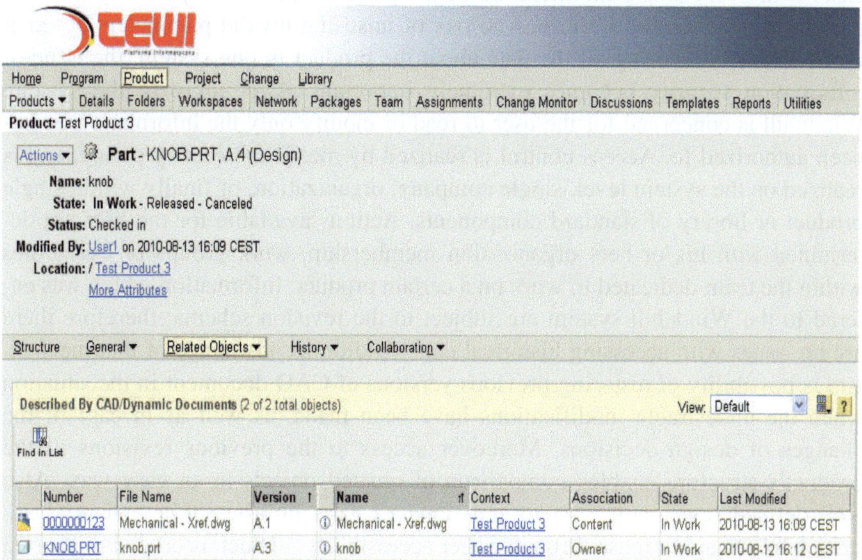

Fig. 7 Data from AutoCAD and Pro/Engineer

coexistence of many various representations is bounded with the constraint of making one of them primary among the other to propagate the attributes. CAD parts which will be creating the structure maybe defined as a 150% Assembly, meaning that the amount of parts will be bigger then needed to create the product. Still though there will be a need to define the primary configuration which will have variants and options possible to use – that provides a lot of flexibility to the engineers.

Assuming that the primary representation in the "bottom-up" method of design are the data created in the Pro/Engineer program, entering data into the system can be accompanied with the automatic creation of the product structure, reflecting the structure of CAD data – so the meta data information are updated based on the models from the CAD tool. Designs made in other tools can be referenced if needed to the primary structure - which leaves some flexibility.

4 Advantages Resulting from Usage of TEWI Platform

Integration of the CAD tools with the PLM system, which is enhancing creation of the products, guarantee obtaining a single source of information in the work environment. All the parameters, which are defined by the user and assigned to the component or assembly, may be moved to the meta-data parts composing the structure of the product. That provides access to meta-data defined information even to users who do not have the possibility of opening CAD documents using the tool dedicated to it. Beside the attributes known as meta-data, also the visualization files are stored, and all the information is being synchronized on a current basis with every single modification of the project data, after uploading them on the server. Due to above, there is no risk of misusing invalid parameters by other users. In case of storing all the data about the product in one system, the issue of information security becomes extremely important. After login to the system, Windchill is concerned for the user to read or modify only the information he has been authorized to. Access control is realized by means of security policy, and is realized on the system level, single company, organization, or finally within single product or library of standard components. Actions available for the user are determined with his or hers organization membership, work groups or even roles within the team dedicated to work on a certain product. Information which was entered to the Windchill system are subject to the revision schema, therefore there are no issues with accessing historical data. Additional advantage of that mechanism is possibility of restoring previous versions of CAD document in the situation when the undesirable modifications have been made, as well as in case of the changes of design decisions. Moreover access to the previous revisions of the product's structure enables comparison of created objects in an easy way. Also well defined revisioning system and standard parts libraries offer enhanced approach to managing reuse through easier accessibility – which is crucial to shorten product development lifecycle.

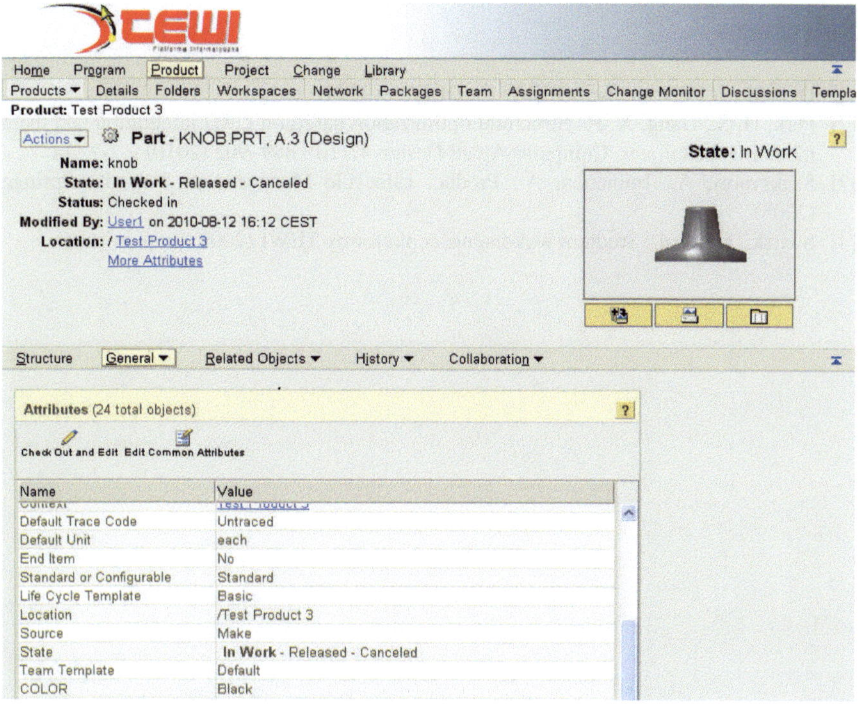

Fig. 8 Coherent and complex source of information.

The main purpose of the arguments presented in this papers is to make the recipient realize the wide possibilities of Windchill system in the domain of coordinating operations on CAD documents and it's superiority over traditional techniques realized by storing data on the file server, or exchanging them via electronic mail or FTP server. PLM system supports creating and designing products, providing not only the platform to exchange the data, but also helps out the in the communication between the persons working on every single phase of product lifecycle. The system provides communication in appropriate context by joining and synchronizing all data and production activities, starting with the design and finishing with service delivery. Ensuring the integrity of the data is strictly connected with the real increase in efficiency of work with the product and shortening the time of launching it to the market. The above is possible thanks to obtaining the higher quality of components of the products, reduction the amount of necessary corrections and early detection of incorrectness.

In today's world, when several months of delay in launching a new product to the market may lead to lower demand caused by though competition there is no place for additional issues like lack of design reuse, weak change management not adequate configuration management. The benefits of integrating CAD tools with PLM systems are becoming a factor, which everybody has to take into consideration to eliminate the risks.

Project donated by UE funds for science, years 2009-2010.

References

[1] Day, M.: Is PLM the New PDM? CADDigest (2004)
[2] Park, H.-S., Dang, X.-P.: Structural optimization based on CAD integration and meta-modeling techniques. Computer-Aided Design 42(10), 889–902 (2010)
[3] Saaksvuori, A., Immonen, A.: Product Lifecycle Management, 3rd edn. Springer (2008)
[4] Świrski, K., et al.: Studium wykonalności platformy TEWI (2009) (unpublished)

Access Control Management Using Extended RBAC Model

Aneta Poniszewska-Maranda

Abstract. The protection of data in information systems against improper disclosure or modification is an important requirement of each system. Nowadays the information systems of enterprises are more and more open, more information is easy accessible for the users, so it is important to better protect the confidential information. The paper presents a software tool for managing the security of enterprise information system on the access control level from the point of view of security administrator. The security administrator who is responsible for information system security will obtain a tool that facilitate the management of one of security aspects, namely the management of access control of users to data stored in a system. However, the application developers can also use such tool to define the access control rules for application elements created on the conception level.

1 Introduction

The protection of data in information systems against improper disclosure or modification is an important requirement of each system. Nowadays the information systems of enterprises are more and more open, more information is easy accessible for the users, so it is important to better protect the confidential information.

Two types of actors cooperate in design process of an information system and its associated security scheme: the information system developer who knows the specifications of information system that need to be realized and, from the other side, the security administrator who has the knowledge of general security constraints that should be respected on the enterprise level. These two actors should cooperate with each other to create the secure information system using the security policies and protocols chosen for the particular company [5,6].

The paper presents a software tool for managing the security of enterprise information system on the access control level from the point of view of security administrator. The security administrator who is responsible for information system security will obtain a tool that facilitate the management of one of security aspects, namely

Aneta Poniszewska-Maranda
Institute of Information Technology, Technical University of Lodz
Wolczanska 215, 90-924 Lodz, Poland
e-mail: anetap@ics.p.lodz.pl

P. Lipiński, K. Świrski (Eds.): Towards Modern Collaborative Knowl. Sharing Syst., SCI 401, pp. 23–33.
springerlink.com

the management of access control of users to data stored in a system. However, the application developers can also use such tool to define the access control rules for application elements created on the conception level.

The created tool used the access control models for definition of security rules and principles in the approach of logical security of information systems. It is possible to distinguish two main categories of access control policies: discretionary security policy and mandatory (non-discretionary) security policy. Many access control models were defined in the field of these two categories. The access control models were created to enforce the rules and objectives of an established security policy and to dictate how the subjects can access the objects. The most common access control models are: Discretionary Access Control (DAC) model, Mandatory Access Control (MAC) model, Role-based Access Control (RBAC) model, extended Role-based Access Control (eRBAC) model, Usage Control (UCON) model. The tool presented in the paper allows the management of access control in security of information system using the extended RBAC model [4] that provides the developers more flexibility and complex vie of the organization security.

The paper is structured as follows: the first part presents the existing traditional access control models. The second part deals with our proposition of a tool – platform for access control management in field of information systems.

2 Access Control Policies and Models

The information system is the set of components, different form the point of view of their usage, e.g. application, databases, database management systems that cooperate with each other in order to obtain the defined purposes of such system. The information systems consist of information technology, communication, management, exchanging information or decision making.

Access control regulates what a user can do directly and what the programs executed on behalf of the user are allowed to do. Access control is a significant part of every information system. The objective of access control is protecting the system resources from any undesired user access [1]. The access control in information systems is responsible for ensuring that all direct accesses to the system objects occur according to the modes and principles defined by protection policies. An access control system contains: **subjects** (i.e. users, processes) who access **objects** (i.e. data, programs) through **operations** (i.e. read, write, run). It consists of two components [1,2]:

- set of access policies and access principles: information is stored in a system and it is accessible by the access modes and
- set of control procedures (security mechanisms) that check access requests in agreement with the principles and rules; these access requests may be allowed, denied or modified, filtering out unauthorized data.

The highest complexity of access control is usually found on the application level. During the definition and creation of access control rules three types of controls should be considered [7]:

- access control policies,
- access control models,
- access control mechanisms.

There exists a distinction between the policies and mechanisms in access control systems. The policies are high-level conceptions, which determine how accesses are controlled and how the access decisions are determined. The mechanisms are low-level software and hardware functions, which can be configured to implement a policy. The security policies of a system generally express the fundamental choices taken by an organization to protect the data. The security policies define the principles on which access is granted or denied.

Access control policies state if and how system subjects and objects can be grouped in order to share access modes, according to given authorizations and rules. The policies state if and how access rights can be transferred. Grouping or classifying users who hold some privileges or resources that share common protection requirements simplifies the specification of the security policies and the implementation of security mechanisms [1,3,7].

There is a numerous amount of various access control models, starting with three most common as Discretionary Access Control (DAC) model, Mandatory Access Control (MAC) model, Role-Based Access Control (RBAC) model and ending with more exotic ones, such as: Attribute-Based Access Control model, Type-Based Distributed Access Control model, Temporal Role-Based Access Control model, Purpose-Based Access Control model, Usage Control model.

Discretionary security model manages the users' access to information according to the user identification and on the rules defined for every user (subject) and object in a system. For each subject and object in a system there are authorization rules that define the access modes of the subject on the object. The access modes: read, write and execute, are verified for each user and for each object [1,2]. "Discretionary" means that users are allowed to grant and revoke access rights on particular objects. This implies decentralization of administration control through ownership. Granting access in the discretionary access control model is transitive. When one has been granted an access, he has the right to distribute that access among any number of other users. DAC systems are used in environments that do not require the structure and higher level of protection that mandatory access control models provide and enforce. Operating systems must be built differently depending upon whether they are going to provide DAC or MAC functionality.

Mandatory (non-discretionary) security model manages the access to the data according to classification of the subjects and objects in a system. Each user and each object of a system are assigned to specific security levels. The access to data is limited by the definition of security classes. Subjects and objects in a system are related to security classes, and the access of a subject to an object is granted if the relation between the classes of the subject and object is verified [1,3]. MAC model do not leave access decisions up to the data owner. Every object in a system has an assigned security label. The security label contains the information about the object's classification (confidential, secret, top secret). Systems implementing MAC models are used in highly secured environments, such as military or government organizations.

Role-Based Access Control model - RBAC model regulates the access of users to the information on the basis of the activities that the users perform in a system. This model requires identification of roles in a system. The role can represent competency to do a specific task and it can embody the authority and responsibility. The permissions are associated with roles and the users are assigned to appropriate roles. The roles are created for various job functions in an organization and the users are assigned roles, based on their responsibilities and qualifications. The user playing a role is allowed to execute all accesses for which the role is authorized. The user can take different roles on different occasions and also several users can play the same role, perhaps simultaneously [1,3].

RBAC model provides support for several important security principles (notably least privilege, privilege abstraction and separation of duties), but does not dictate how these should be put into practice [4]. Role-based access control models is highly effective especially in large companies that have high turnover rates due to the fact of allowance for the administrator to simply place the new employees into the roles instead of creating the new permissions for each new person who joins the company and needs to be included in the system.

Extended RBAC (eRBAC) model - the complexity of organizations gave rise to the idea of extending the standard RBAC model to ensure a finer decomposition of the responsibilities in an enterprise. To this aim, the notion of a function has been introduced. Since a person employed in an enterprise may have many responsibilities in it, he may be attached to an entire set of roles. Each role defined in the extended RBAC model makes it possible to realize a specific task associated with the enterprise process. At the same time, every role can contain many functions that a user can take, and therefore it is possible to choose functions in the system that are necessary for a given role [4,5]. Therefore, the extended RBAC model is the classical RBAC model extended by the addition of some elements, i.e. function, object, method, class, operation, to express more complex elements of an information system that are secured by these model (Fig. 1).

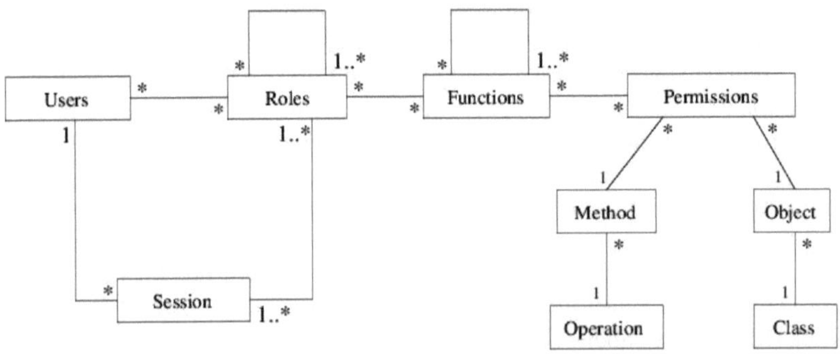

Fig. 1 Extended RBAC model

3 Platform for Management of Access Control

The platform for management the security of information system on access control level is designated first of all for security administrator. This platform uses different concepts presented in [4] and [6].

Two types of actors cooperate in design process of an information system and its associated security scheme: the information system developer who knows the requirements and specifications of information system that need to be realized and, from the other side, the security administrator who has the knowledge of general security rules and constraints that should be respected on the enterprise level (Fig. 2).

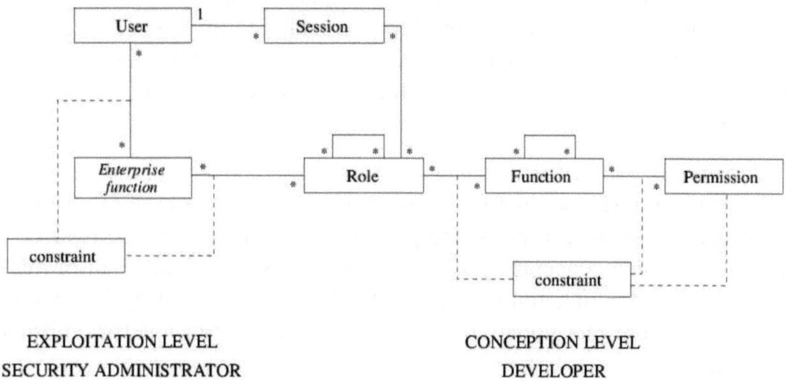

Fig. 2 Partition of responsibilities of two actors - application developer and security administrator

The application developer realizes the stage of design. He defines the application model that contains all the elements expressing the user needs. Developer generates from this model the sets of following elements: roles, functions, permissions and constraints. These sets of elements should be presented to security administrator in a useful and legible form.

In the second stage, i.e. during the exploitation phase of information system, the security administrator assigns the identified roles to the users based on their responsibilities and on global security policy defined for the enterprise.

The most important task in application and realization of RBAC model or eRBAC model is the proper identification of set of roles that a user can take in the application or in whole information system. It is important to remember about the separation of role management between:

- developer of information system who identifies and defines the roles of an application and
- security administrator who assigns the roles to users of the application and/or information system.

The creation process of security scheme in an information system using the RBAC or eRBAC model is proposed as follows:

- Application developer creates the system application or a set of system applications using the UML (Unified Modeling Language) concepts. UML is used to define the application Model containing all elements that express the needs of the users.
- Application developer initiates the process of user profile creation (e.g. role engineering) [4] based on the security rules concerning this application.
- The application's Model created by the developer is translated to the concepts of access control models (e.g. RBAC or eRBAC) based on the connection of UML concepts with the concepts of RBAC/eRBAC model [4,6]. Also the process of user profile creation is finished on the developer level.
- Security administrator receives the Model containing the lists of access control elements which are presented in a special form, e.g. in XML files. The administrator finishes the process of user profile creation using the rules of global security policy.

The main purpose of the management tool is the cooperation between the application/system developer and the security administrator to assure the coherence of information system security on the global level. This task can be realized with the use of presented concepts of access control [4,6]. The final result of this tool will be the validation of activities realized by the developer and by the administrator who will assure the global coherence on access control level in information system (Fig. 3).

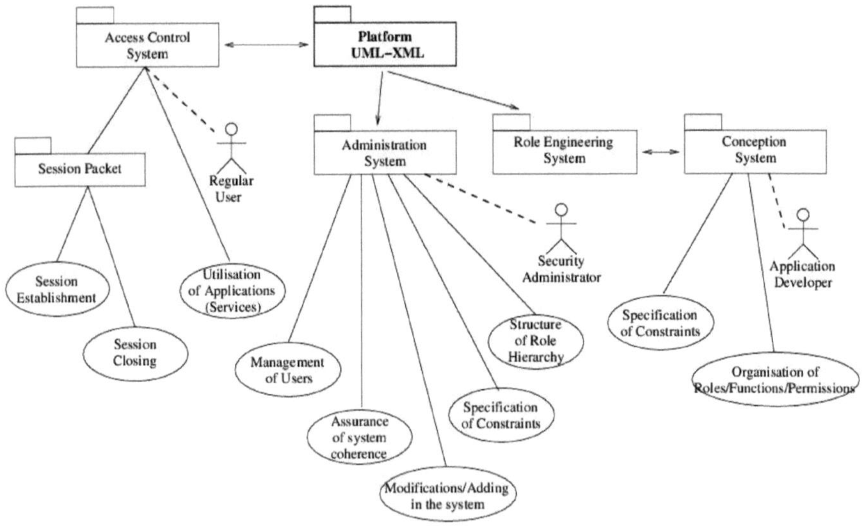

Fig. 3 Platform functionality

The tool for management of access control in the information systems was dedicated first of all to the security administrator who has the knowledge of general security rules, privileges and constraints that should be respected on the enterprise level. The application/system developer who knows the specifications of information system that need to be realized can also use the tool [4].

The main technologies of software engineering used to realize the platform are:

- CASE tool to create the logical model of an application with the use of UML concepts,
- XML language to exchange the information came from different UML diagrams, assuring the independent formalism and
- object-oriented language, e.g. Java language that is independent of the destination system.

The process of role production (i.e. role engineering) based on extended RBAC model uses the concepts presented above and widely in [4] and [6]. Two languages – the UML and XML are used in course of process of role engineering. Figure 4 presents the stages of role production.

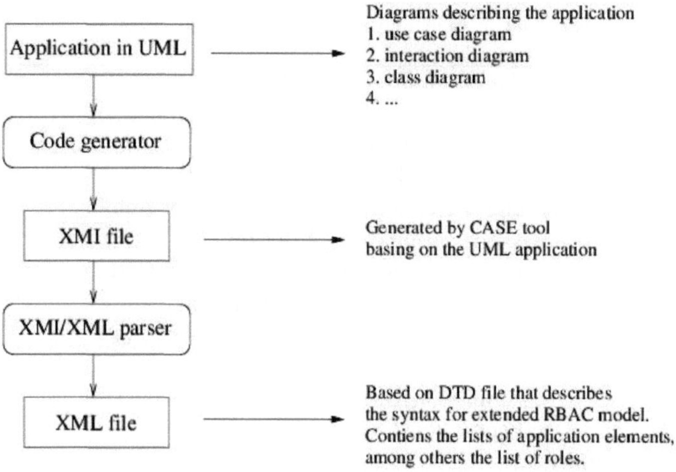

Fig. 4 Functionality stages of XMI/XML parser

The first phase of role production is the creation of roles and definition of set of roles. This stage is realized basing on associations between the eRBAC concepts and UML concepts [5]. The specification of access control on level of concept schema of information system security was proposed using the UML. The algorithm of role construction of an information system was specified using UML meta-model. The stages of this algorithm are as follows (Table 1):

- assignment a set of permissions to a function (an use case in conceptual logical model) – Table 2 and
- assignment a set of functions to a role (an actor in conceptual logical model) – Table 3.

Table 1 Algorithm for construction of roles in role engineering process

```
Construction of Roles for the Model
Begin
   Assignment of Permissions to Functions for the Model
   Assignment of Functions to Roles for the Model
End
```

Table 2 First stage of role creation algorithm

```
Assignment of Permissions to Functions for the Model
Begin
   setUseCases = search Use Cases for the Model
   for each UseCase ∈ setUseCase do
      setPermissions = creation set of Permissions for UseCase
   done
End
```

Table 3 Second stage of role creation algorithm

```
Assignment of Functions to Roles for the Model
Begin
   setActors = search Actors for the Model
   for each Actor ∈ setActors do
     setFunctionsActor = creation set of Functions for Actor
     for each UseCase_i ∈ setFunctionsActor do
        setRelationUseCase_i = search Relations of UseCase_i
        for each UseCase_j ∈ setRelationUseCase_i do
           if (relation(UseCase_i, UseCase_j) is type <<include>>)
             then
                setPermissionsUseCase_i = setPermissionsUseCase_i ∪
                setPermissionsUseCase_j
           endIf
           if (relation(UseCase_i, UseCase_j) is type <<extend>>)
             then
                setFunctionsActor = setFunctionsActor ∪ UseCase_j
           endIfi
        done
     done
     inheritanceSetActors = search Generalization Relations of Actor
     for each actor_j ∈ inheritanceSetActors do
        setFunctionsActor = setFunctionsActor ∪ setFunctionsActor_j
     done
   done
End
```

The second phase of role production is the creation of XMI file. This file can be generated automatically for the application created in UML, using the CASE tool. Next, the created parser of XML files uses the UML meta-model to analyze the received XMI files. XMI/XML parser is used to receive a tree of elements containing all the elements corresponding to extended RBAC model. The functionality of XMI/XML parser is described by the following tasks:

- passage of XMI file and identification of elements existing in two types of diagrams (i.e. use case diagram and interaction diagram): actors, use cases, their relations, classes, objects and messages sent between the objects,
- definition of sets of identified elements,
- identification of these elements as the elements characteristic for the concepts of extended RBAC model,
- determination of sets of elements of particular application as the sets of elements characteristic for extended RBAC model,
- generation of XML file that contains the sets of determined elements.

The user of this parser receives the file in XML format and it contains all elements and their proprieties of application designed in UML. These elements are next used in the integration of eRBAC model. Received XML file contains the elements of extended RBAC model: roles, functions, permission, objects and methods. The parser generates automatically the XML file. The formal description of the elements of this file exists in DTD file that main elements have the following form:

```
< !ELEMENT eRBAC(role+, function+, permission+, method+, object+, operation+, class+>

<!ELEMENT role(name, refRole*, refFunction+)>
<!ELEMENT function(name, refRole+, refFunction*, refPermission+)>
<!ELEMENT permission(refMethod+, refObject+, refFunction+)>
<!ELEMENT method(name, refPermission+, refOperation+, attribut*)>
<!ELEMENT object(name, refPermission*, refClass+, baseClass)>
<!ELEMENT operation(name, refMethod*)>
<!ELEMENT class(name, refObject})>
```

The result document is the product of developer job on access control level of an information system. It contains the elements of an application created by the developer and translated into concepts of extended RBAC model. Next, this document is given the security administrator who manages the security of the whole system. He should integrate the new elements, e.g. the new application, in the system on the security level. The XML document allows realizing of such integration independent of the environment. It describes the elements of new application on the RBAC level used by the administrator of access control. The security administrator uses the XML document to define the rights of users in the system.

The security administrator should realize the following tasks for each new element, e.g. new application, added to the information system:

- define the security constraints on the administrator level,
- add the new roles of a new application into role hierarchy of the system,
- add the security constraints of a new application into the set of system constraints,
- add the relations between the elements of a new application and the relations between system elements and new application elements into the set of system relations according to defined security constraints,
- define the user profiles for each user who will use the application.

The platform should realize the validation of integration of application/system developer job and security administrator job. The validation should guarantee the global coherence of the information system after integration of a new application on access control level.

The minimal set of tool functionality should provide the security administrator the possibility of managing the elements that are characteristic for extended role-based access control model. First of all, it should be possible to manage the roles defined in the system obtained from the list of system roles. Adding new roles, modifying already existing ones or removing the particular roles can modify such list. The name of the role can be changed and the role can be grouped in hierarchy of roles by setting the parent roles.

The administrator can also manage the system users, defining the user profiles. He can add the new users to the system, change the names, passwords and assign the users to the existing roles. There is also the possibility of deleting some of users from the list of users on the access control level.

4 Conclusions

The administration tool presented in the paper represents a software tool to manage the logic security of enterprise information system from the point of view of application/system developer and from the point of view of security administrator. Particularly, the security administrator, who is responsible for the information system security, can obtain the tool to facilitate the management of one of security aspects, namely the management of access control of users to data stored in a system.

The main purpose of the tool it to allow the management of logical security in aspect of one of the access control models, i.e. extended RBAC model. It can be used to define the user profiles in the information system. The process of user profile creation is the main stage in definition of security policy of an information system on the access control level. It is realized by two actors, system developer and security administrator, that cooperate with each other to assure the global coherence on logic security level.

References

[1] Castano, S., Fugini, M., Martella, G., Samarati, P.: Database Security. ACM Press, Addison-Wesley (1994)

[2] Sandhu, R., Munawer, Q.: How to do Discretionary Access Control Using Roles. In: Proceeding of 3rd ACM Workshop on Role-Based Access Control (1998)

[3] Ferraiolo, D., Sandhu, R., Gavrila, S., Kuhn, D., Chandramouli, R.: Proposed NIST Role-Based Access Control. ACM Transactions on Information and Systems Security TISSEC 4(3) (2001)

[4] Goncalves, G., Poniszewska-Maranda, A.: Role engineering: from design to evaluation of security schemas. Journal of Systems and Software 81 (2008)

[5] Poniszewska-Maranda, A., Goncalves, G., Hemery, F.: Representation of Extended RBAC Model Using UML Language. In: Vojtáš, P., Bieliková, M., Charron-Bost, B., Sýkora, O. (eds.) SOFSEM 2005. LNCS, vol. 3381, pp. 413–417. Springer, Heidelberg (2005)

[6] Poniszewska-Marańda, A.: Conception Approach of Access Control in Heterogeneous Information Systems using UML. Journal of Telecommunication Systems 44 (2010)

[7] Sandhu, R., Samarati, P.: Access control: Principles and practice. IEEE Communication 32(9), 40–48 (1994)

Evaluation of DocBook 5.0 in Documenting Research Applications

Kamil Kowalewski, Marek Parfieniuk, Oskar Świda, and Marcin Kazberuk

Abstract. This chapter presents practical information about DocBook 5.0 which was obtained by examining the usefulness of this technology for documenting a scientific application for the TEWI platform: a compiler for a domain-specific language for implementing transforms and filter banks. Results of the study may be useful for people who need to choose a format and tools for creating and organizing scientific/technical documents or teaching materials.

1 Introduction

Creating and maintaining documentation is a vital issue in software engineering. Ensuring high quality and validity of both project and end-user documentation requires hard work, and thus time and money, especially when the application to be documented is still intensively developed. Another important problem is to effortlessly present the same content in various formats (print, Web, integrated help), each of which can just as well be preferred by application users or business partners of a software producer. Finally, all documents should look uniform and be thoroughly indexed to facilitate their searching.

The aforementioned problems can be largely solved by creating documents using text structuring, i.e. by supporting content with metadata that describe its semantics and allow it to be separated from formatting. Both these postulates are satisfied by the DocBook technology [1, 2], which is based on the XML language and is used by both the open-source community and IT giants.

However, most of software are not of a scientific nature and thus can be documented without using complex mathematical expressions or numerous literature references. Thus, in spite of the optimistic tone of [1], it is not obvious if DocBook 5.0 is useful in the case of research applications, which are based on complex algorithms and knowledge distributed among many various publications and presented using mainly mathematical notations. In addition, as a technology developed mainly by English-speaking designers and developers, DocBook can be supposed to have difficulties with processing languages such as Polish or Russian.

Kamil Kowalewski · Marek Parfieniuk · Oskar Świda · Marcin Kazberuk
Faculty of Computer Science, Bialystok University of Technology
Wiejska 45a, 15-351 Bialystok, Poland
e-mail: m.parfieniuk@pb.edu.pl, o.swida@pb.edu.pl

P. Lipiński, K. Świrski (Eds.): Towards Modern Collaborative Knowl. Sharing Syst., SCI 401, pp. 35–48.
springerlink.com

Implementing a scientific application for the TEWI platform, a compiler for a domain-specific language for implementing transforms and filter banks, the authors of this study needed a technology for preparing documentation. Knowing the capabilities of LaTeX and word processors, they decided to investigate DocBook 5.0 as an alternative, so as to be able to select the tool best suited to their needs.

This chapter presents practical information which was obtained from experiments with DocBook. It should allow readers who do not have an experience with this technology to learn the basics and to consider employing it in their projects.

2 General Characterization of the DocBook Technology

2.1 DocBook as an XML Application

DocBook 5.0 is an XML application, that is, it is a specialized markup language (set of tags, also called elements), which is defined as a Relax NG schema. Both DTD (Document Type Definition) and XML Schema specifications are also provided, but their use is not recommended because these technologies do not allow some constraints to be expressed.

DocBook was invented to help with editing documentations, books, articles, and even slides. Although the technology was developed by IT companies, it was designed to be useful in various fields of science and engineering. Its versatility has been achieved by developing hundreds of elements, which can potentially exist in technical documents of different kinds. In [1], the DocBook tags are grouped into the categories presented in Table 1.

DocBook describes all elements that can be used to create a document and defines rules for their use. The author must strictly structure documents, but, in return, he receives the possibilities of exchanging content and automatic formatting. It is possible to extend DocBook with MathML and SVG, XML-based markup languages for describing mathematical expressions and vector graphics, respectively. On the other hand, one can limit the set of tags, if this simplifies document editing or processing without affecting the desired functionality.

DocBook documents are also XML files, as shown in Figure 1. They are not suitable for presentation because tags obscure text. A presentation form results from transforming their contents, which can be done using three technologies: XSLT, CSS, and XQuery. The first of them is most commonly used, whose essence is explained in Figure 2.

XSLT (Extensible Stylesheet Language Transformations) is an XML-based language for describing transformations of XML documents into other formats, not necessarily based on XML, including clean text. The XSLT files are called stylesheets. They are difficult to write or modify for non-specialists, but a rich set of stylesheets for formatting DocBook documents are provided online [3].

Transformations are performed by programs called XSLT processors, like "xsltproc", "Saxon", and "Xalan". The possibility of processing documents using standard XML tools is a very important advantage of the DocBook technology.

It is also noteworthy that processing of DocBook data can go beyond formatting. In [4], DocBook is used in remote education for the purpose of content personalization and profiling in accordance with student proficiency.

2.2 Origin and Evolution of DocBook

DocBook was created in 1991 by Hal Computer Systems and O'Reilly publishing house for the purpose of exchanging documents between these two companies. Originally it was an SGML (Standard Generalized Markup Language) application, and formatting was based on DSSSL (Document Style Semantics and Specification Language). Over the years, the popularity of DocBook grew, and more companies wanted to be involved in its development. This resulted in establishing standardization committees: the Davenport Group, in 1994, and then, in 1998, the DocBook Technical Committee under OASIS (Organization for the Advancement of Structured Information Standards).

Table 2 shows the milestones in the evolution of DocBook. Unlike the previous versions, DocBook 5.0 is not fully backward compatible. This was necessary to meet user demands and to prepare a basis for future versions. Currently, version 6 is developed.

Table 1 Main categories of DocBook elements

Category	Comment
Sets	The "set" element, which can contain a number of books, articles, documentations, or other sets.
Books	The "book" element, which is responsible for a logical schema of a document, being able to contain dedication, navigational elements (tables of contents, lists of figures, etc.), divisions of a book into parts, and other components.
Components	Chapters, book parts, and similar items: appendix, glossary, bibliography, etc.
Sections	Smaller parts of components.
Metadata	Describe documents and their parts mainly from a bibliographical point of view (abstract, author, publication date, version history).
Block elements	Represented by paragraphs, related to line breaking: lists, comments, listings, screenshots, drawings, tables, paragraphs, equations, multimedia objects, etc.
Inline elements	Allowed to appear in text: abbreviations, acronyms, emphasis, cross-references, inline equations, elements of graphical user interface (GUI), programming language syntax, operating system parameters.

```
<?xml version="1.0" encoding="UTF-8"?>
<?oxygen RNGSchema=http://www.oasis-open.org/docbook/xml/5.0/rng/docbook.rng
  type="xml"?>
<book xmlns="http://docbook.org/ns/docbook"
      xmlns:xlink="http://www.w3.org/1999/xlink" version="5.0">
  <info>
    <title>Opracowanie metod integracji kompilatora TDL z TEWI</title>
    <subtitle>Projekt TEWI</subtitle>
      <authorgroup>
        <author><personname>
            <firstname>Kamil</firstname><surname>Kowalewski</surname>
        </personname></author>
      </authorgroup>
    </info>
    <preface>
      <title>Cel dokumentu</title>
      <para>
         Dokument powstał w ramach umowy o dzieło nr
         <emphasis>TEWI/WI/D/4/10</emphasis> polegające na
         <emphasis role="italic">opracowaniu i wdrożeniu metod wykorzystania
         platformy TEWI do dystrybuowania dokumentacji kompilatora
         TDL</emphasis>.
       </para>
    </preface>
    <xi:include href="parts/chapters/szablon.xml"
      xmlns:xi="http://www.w3.org/2001/XInclude" />
    <xi:include href="parts/chapters/dokumentowanie.xml"
      xmlns:xi="http://www.w3.org/2001/XInclude" />
    <xi:include href="parts/chapters/instalacja.xml"
      xmlns:xi="http://www.w3.org/2001/XInclude" />
    <appendix>
      <title>Zawartość płyty CD</title>
      <para>
        Opracowane materiały, pliki źródłowe oraz dokumenty stanowiące
        załącznik do niniejszego dokumentu znajdują się na dołączonej
        płycie CD. Zawartość płyty opisuje tabela
        <xref linkend="cd" xrefstyle="select: labelnumber"/>.
      </para>
    </appendix>
    <xi:include href="parts/common/bibliography.xml"
      xmlns:xi="http://www.w3.org/2001/XInclude" />
</book>
```

Fig. 1 Example of a DocBook file

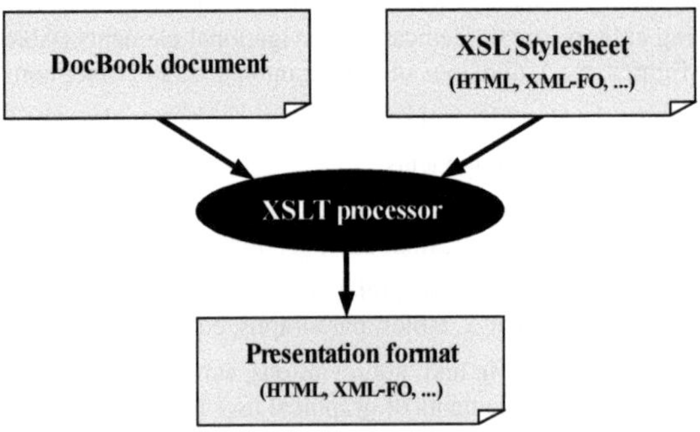

Fig. 2 XSLT-based processing of DocBook documents

Table 2 Evolution of the DocBook standard

Year	Version	Base	Organization
1991	b.d	SGML	HCS / O'Reilly
1994	1.2.2.	SGML	Davenport Group
1999	3.1	SGML	DTC OASIS
2001	4.1.2	SGML/XML	DTC OASIS
2005	5.0 beta	XML	DTC OASIS
2006	4.5	SGML/XML	DTC OASIS
2009	5.0	XML	DTC OASIS

It is noteworthy that OASIS is a consortium which creates, develops and maintains standards for information exchange. It was founded in 1993 under the name "SGML Open" in order to promote and support technologies based on SGML. Its name was changed in 1998 with the aim of expressing the interest of OASIS in XML and other related standards.

The consortium currently holds about 60 standards for network services, data exchange, security and e-business. It brings together more than 5000 members, which represent over 600 organizations and more than 100 countries. Half of the members are with companies that create technologies, a third part is with companies that use new technologies, and the remaining ones represent government organizations and universities. In particular, among the participating companies are Boeing, IBM, Microsoft, Google, Intel, HP, Cisco, Nokia and Oracle. An example of a Polish participant in OASIS is the Polish Association of Translation Offices (PSBT).

Despite its commercial origins, DocBook is also the standard for creating software documentation in the open-source world, being used in such well-known projects as Linux Documentation Project (LDP), GNOME, or GTK+.

3 Documenting Scientific Applications and Research Results Using DocBook

3.1 Inserting Equations

Displayed mathematical formulas are inserted into DocBook documents using the “equation” tag, which is also responsible for equation numbering. A title can be assigned to a formula, which appears above or below it. In the absence of the title, the equation number is reduced and displayed on the right-hand side.

DocBook itself provides no way to describe formulas. In some publications, e.g. in [1], the "subscript" and "superscript" tags are regarded as tools for writing equations, but actual handling mathematical notations requires employing much more powerful tools.

The recommended solution, which ensures consistency of XML documents, is to extend DocBook with the MathML tags. MathML is a markup language recommended by W3C to publish mathematical expressions in hypertext documents [5]. As illustrated in Figure 3, it is extremely verbose and difficult to read even in the case of simple formulas, so typing tags manually makes little sense. An equation editor is necessary that supports MathML export, such as MathType.

Figure 3 also shows that LaTeX outperforms MathML in terms of conciseness. As shown in Section 3.5, it is possible to embed LaTeX code in DocBook documents, but this requires complicating processing.

(a)

$$\frac{-b \pm \sqrt{b^2 - 4ac}}{2a}$$

(b)

```
<math display='block'>
  <semantics>
    <mrow>
      <mfrac>
        <mrow>
          <mo>&#x2212;</mo>
          <mi>b</mi>
          <mo>&#x00B1;</mo>
          <msqrt>
            <mrow>
              <msup>
                <mi>b</mi>
                <mn>2</mn>
              </msup>
              <mo>&#x2212;</mo>
              <mn>4</mn>
              <mi>a</mi>
              <mi>c</mi>
            </mrow>
          </msqrt>
        </mrow>
        <mrow>
          <mn>2</mn>
          <mi>a</mi>
        </mrow>
      </mfrac>
    </mrow>
  </semantics>
</math>
```

(c)

```
\[
  \frac{
    -b \pm \sqrt{{b^2} - 4ac}
  }{
    2a
  }
\]
```

Fig. 3 Simple equation and its representations in MathML and LaTeX

3.2 Inserting Tables

DocBook allows tables to be created using two models: CALS and HTML. The CALS table model is recommended by OASIS as a standard for representing tabular data in XML documents. The HTML model is consistent with the usual way of creating tables for the Web. The example of using the CALS model is shown in Figure 4, which allows the reader to compare the CALS and HTML approaches.

After testing different DocBook tools, the authors have concluded that it is better to use the CALS model, because some programs (e.g. "dblatex", see Section 3.5) do not support the HTML model. Apart from this, no significant problems with table processing have been encountered.

3.3 Inserting Graphics

Inserting graphics into a DocBook document requires only indicating the image file, whose format does not matter: even a multimedia clip can serve as a source. Figure 5 presents the appropriate tags and shows how to refer to the image from other parts of the document.

Actual positioning of graphics elements is closely related to document formatting, so its results highly depend on the used XSLT processor and selected output format. For example, if a target format lacks the concept of page breaking, as in the case of HTML, then one can expect problems with positioning drawings that are not anchored to text, so-called floats. By contrast, embedding a video clip makes little sense in the case of printed output. Rather paradoxically, in both cases, the documents are valid from the point of view of the DocBook syntax.

It is possible to directly embed SVG vector graphics into DocBook documents. However, the correctness of visualization of such images depends strongly on the processor used to format documents.

3.4 Support for Non-english Languages and Multilingual Documents

Writing a fragment of a DocBook document in Polish or German is straightforward, provided that the appropriate encoding is set in the XML prolog and that the file is edited using an editor that supports this encoding. In particular, the UTF-8 encoding allows multilingual documents to be processed. In such documents, the language can be specified for elements by setting the "xml:lang" attribute, which allows editing software to use the appropriate dictionary for spell-checking.

Formatting documents with non-English characters is more troublesome than editing them, as it requires reconfiguring both XSLT tools and stylesheets. In particular, seeing national characters in PDF files requires Apache FOP to be switched so as to use appropriate fonts, and what is more, these fonts must be specifically prepared.

(a)

```
<table><title>Tabela</title>
  <tgroup cols="3" rowsep="1" colsep="1">
    <colspec colname="c1"/>
    <colspec colname="c2"/>
    <colspec colname="c3"/>
    <thead>
      <row>
        <entry align="center">
          Nagłówek 1
        </entry>
        <entry namest="c2" nameend="c3" align="center">
          Polaczony nagłówek
        </entry>
      </row>
    </thead>
    <tbody>
      <row>
        <entry morerows="1" align="center" valign="middle">
          Złączone dwa wiersze
        </entry>
        <entry>2.1</entry>
        <entry>3.1</entry>
      </row>
      <row>
        <entry>2.2</entry>
        <entry>3.2</entry>
      </row>
    </tbody>
  </tgroup>
</table>
```

(b)

Tabela 6.1. Tabela

Nagłówek 1	**Polaczony nagłówek**	
Złączone dwa wiersze	2.1	3.1
	2.2	3.2

Fig. 4 A CALS-based table description and the result of its formatting

(a)

```
<figure xml:id="obr2">
    <title>Tytuł obrazka</title>
    <mediaobject>
      <imageobject>
        <imagedata fileref="fig/pic1.jpg" scale="50"/>
      </imageobject>
      <caption>
        <para>Podpis pod obrazkiem</para>
      </caption>
  </mediaobject>
</figure>
```

(b)

```
<xref linkend="obr2"/>
```

Fig. 5 Embedding image file in a DocBook document and referring to it

Firstly, for a given TTF file, a metric file must be generated which describes the font in a special XML-based format. The TTFReader application can be used for this purpose, which is a part of Apache FOP distributions and can be invoked from the command line in the way shown in Figure 6a. Secondly, both TTF and metric files need to be registered in the FOP configuration file, “fop-config.xml”, as illustrated in Figure 6b. If the italic and/or bold versions of the same font have to be used, they must be separately registered with different names and appropriate values of the “style” and “weight” attributes. Finally, font names must be assigned to elements of the FO XSL stylesheet, as shown in Figure 6c.

The formatting styles provided with DocBook automatically generate headlines such as "Contents" and fragments of captions such as "Figure". The default language for these document elements is English, so in order to have them translated into Polish, it is necessary to edit the “param.xsl” file of the FO XSL stylesheet, so as to set the “l10n.gentext.default.language” parameter to the "pl" value.

Even though the aforementioned tasks are not very difficult, caring about details makes them quite laborious, especially if multiple fonts are used in a document. Nevertheless, it is possible to create even multilingual documents, if fonts are accessible that support a sufficiently broad set of characters (it is easiest to use a Unicode font). An example is demonstrated in Figure 7.

(a)

```
java -cp fop.jar;avalon-framework-4.2.0.jar;commons-logging-
1.0.4.jar;commons-io-1.3.1.jar;xmlgraphics-commons-1.3.1.jar
org.apache.fop.fonts.apps.TTFReader ARIAL.TTF arial.xml
```

(b)

```
<!-- in fop-config.xml (the FOP configuration file) -->
  ...
<base>..</base>   <!-- root for file references -->
  ...
<fonts>
  ...
  <font metrics-url="fopConfig/fonts/arial.xml"
      kerning="yes" embed-url="fopConfig/fonts/ARIAL.TTF">
    <font-triplet name="Arial" style="normal" weight="normal"/>
  </font>
  ...
</fonts>
```

(c)

```
<!-- in param.xsl (the FO XSL stylesheet) -->
  ...
<xsl:param name="body.font.family">TimesNewRoman</xsl:param>
<xsl:param name="monospace.font.family">CourierNew</xsl:param>
<xsl:param name="sans.font.family">Arial</xsl:param>
```

Fig. 6 Using the Arial Unicode font to format multilingual Docbook documents: (a) generating a metric file, (b) registering it in the Apache FOP configuration, and (c) incorporating the font into a stylesheet

(a)

```
<?xml version="1.0" encoding="UTF-8"?>
<?oxygen RNGSchema="http://www.oasis-
open.org/docbook/xml/5.0/rng/docbook.rng" type="xml"?>
<book xmlns="http://docbook.org/ns/docbook" version="5.0">
  <chapter>
    <title>Multilingual documents</title>
      <sect1>
        <title xml:lang="en">German</title>
        <para xml:lang="de">Eine Frau und ein Mann schließen Ehe,
          weil sie nicht wissen, was sie miteinander machen können.
        </para>
      </sect1>
      <sect1>
        <title xml:lang="en">Polish</title>
        <para xml:lang="pl">Na to, co naprawdę chcemy robić, zawsze
          znajdujemy czas.</para>
      </sect1>
      <sect1>
        <title xml:lang="en">Russian</title>
        <para xml:lang="ru">мы все найти время, чтобы делать то,
        что мы действительно хотим сделать</para>
      </sect1>
    </chapter>
</book>
```

(b)

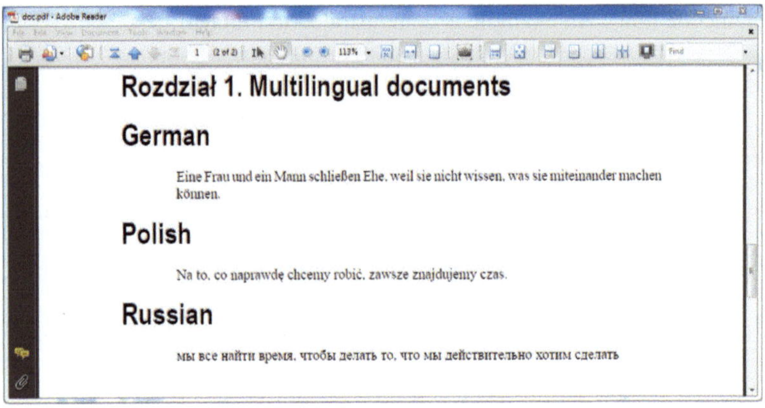

Fig. 7 A multilingual Docbook document: (a) XML source and (b) PDF output

3.5 *Converting Documents from DocBook to LaTeX*

DocBook documents can be converted into LaTeX source files using the "dblatex" tool [6], which was written in Python and runs in two phases. First, it uses XSLT transformations to prepare a LaTeX file. However, the mechanism produces files that contain XML-compatible elements which are undesirable in LaTeX. For example the characters '<', '>', '&' are represented in XML by entities: "<", ">", and "&amp", respectively. Therefore, the second stage follows that eliminates these XML elements.

A great advantage of "dblatex" is that it allows equations in DocBook documents to be written using the LaTeX syntax. Unfortunately, XML tools do not facilitate editing such expressions. A less important yet practical feature of the tool is its ability to automatically transform LaTeX files into PostScript or PDF.

It should be mentioned that there exists a similar tool named "db2latex". It is less convenient as only simple file conversion from Docbook to LaTeX is performed. A user is responsible for further processing of the resulting file, i.e. for running LaTeX tools. Moreover, the development of "db2latex" seems abandoned. In particular, poor documentation is provided, and the recent version (issued in 2004) is not compatible with DocBook 5.0.

A related question is if bibliographic data in the Bibtex format can be converted to DocBook. The answer is positive as tools exist for this purpose. In particular, Bib2DB5 can easily be downloaded and used, but our tests revealed that it cannot recognize some metadata. In such cases, the tool generates output but with incomplete bibliographic information.

It is better to use JabRef, a bibliography manager that supports export to DocBook 4.1.2. Even though in DocBook 5.0, elements related to bibliography have been redefined, files obtainable from JabRef can easily be corrected. It is only necessary to add the <personname> tag between the <author> tag and its children.

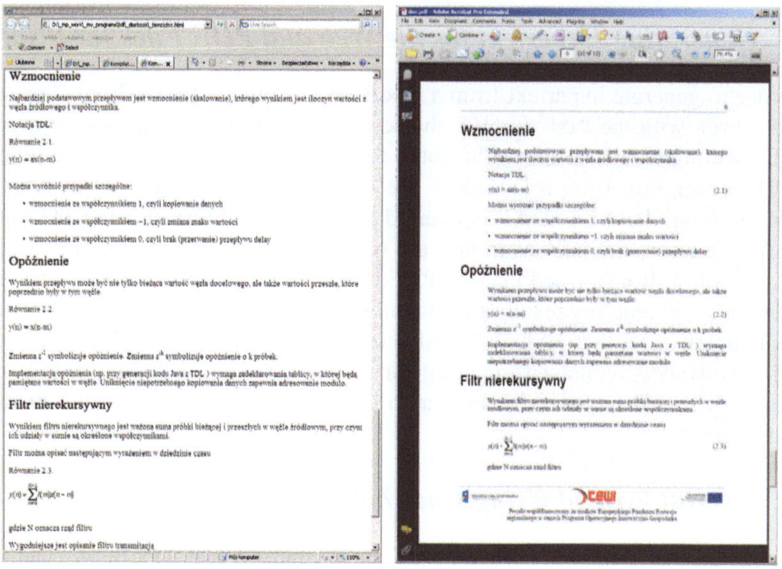

(a) (b)

Fig. 8 Results of transforming a DocBook document into (a) HTML and (b) PDF

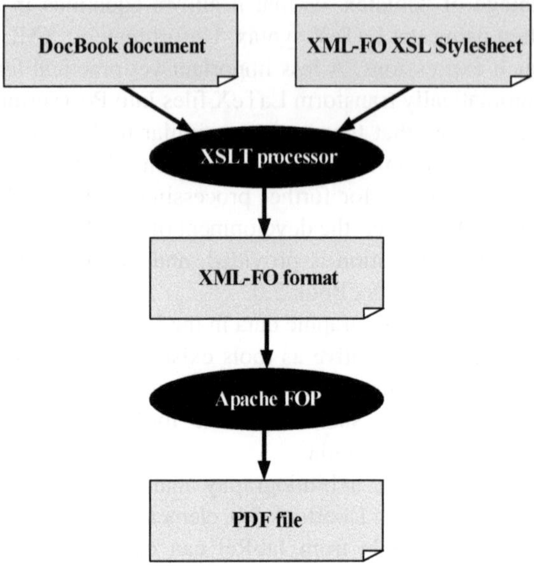

Fig. 9 Conversion of DocBook documents into PDF

3.6 *Generating Hypertext for the Web and PDF Files for Printing*

In order to generate hypertext from a DocBook document, the "xsltproc" program must be run with the "xsl/xhtml/docbook.xsl" style. Figure 8a allows for evaluating formatting results. It should be noted that equations are converted into images.

PDF generation from a DocBook file consists of two stages, as illustrated in Figure 9. First, the document is converted to the XSL-FO format using an ordinary XSLT processor, similarly as in the case of generating HTML. Then, the resulting XSL-FO file is transformed into the PDF format using programs like Apache FOP or RenderX. Exemplary results are shown in Figure 8b.

It is noteworthy that XSL-FO (XSL Formating Object) is an XML-based standard for advanced content formatting for different media, including displays and printers. XSL-FO is generally similar to CSS but has greater potential for document paging.

3.7 *Overall Ease of Using DocBook*

Persons familiar with XML have no difficulty starting with DocBook. The technology seems also accessible for people who created publications using LaTeX or webpages using HTML.

Editing itself is quite simple and enjoyable provided a writer is equipped with a good text editor, or, even better, an editor that supports XML, such as <oXygen/>, which constantly checks document structure, provides document templates for various purposes, and even allows for editing in a WYSIWYG-like mode as illustrated in Figure 10. However, carefully defining element semantics requires more effort than working with word processors as well as with LaTeX. Additionally, in some situations, it can be difficult to precisely and unambiguously define element semantics.

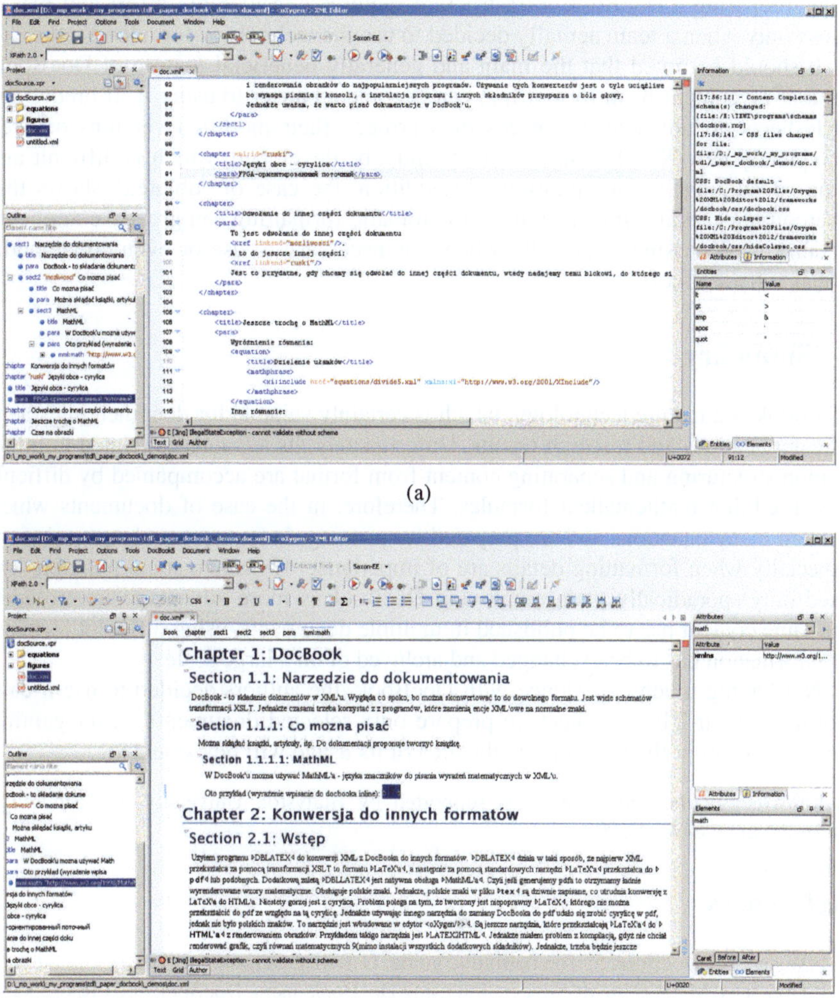

(a)

(b)

Fig. 10 Example of a modern XML editor: <oXygen/> in (a) the text and (b) author modes

It is more troublesome to format documents, which requires running an XSL processor and possibly programs that support it. Running these tools usually requires setting multiple command-line parameters and lengthy file paths. In practice, it is necessary to develop scripts that automate the conversion and/or to use an advanced editor that releases a user from manually invoking external programs.

Preparing a work environment seems also a problem of no small importance. It is necessary to choose several programs for different tasks (DocBook editor, XSL processor, equation editor, etc.), to develop a workflow based on their cooperation, to adjust their settings, and to configure fonts and stylesheets. It is advantageous to designate one of the members of a project team to specialize in this field. Thus preparing a DocBook infrastructure requires considerable effort and makes sense only when a team actually decided to use the technology for a long time.

It should be noted that the main and generally beneficial feature of DocBook, separating content from formatting, can become a barrier to using it. If predefined stylesheets do not meet the needs of a project, their modified versions or even completely new XSLT transformations must be developed, which is difficult and time consuming in real situations. In addition, the case of "dblatex" shows that overcoming certain limits or faults of automatic formatting can greatly complicate document processing, especially when it is necessary to use or even create additional software tools.

4 Summary

DocBook is a mature technology, which is certainly suitable for documenting scientific applications and research results. Unfortunately, its advantages related to information structuring and separating content from format are accompanied by difficulties in editing mathematical formulas. Therefore, in the case of documents which contain many equations and are prepared for printing, Latex seems a better solution, especially when formatting details are of importance. If mathematical notations are used only sporadically, then employing DocBook is more advantageous provided the same content has to be published in multiple documents, using different formats, or information has to be exchanged and archived of on a large scale.

Not having much experience with DocBook, the authors decided to use it cautiously in their TEWI project, to prepare only selected documents. After gaining knowledge about the technology, the area of its usage will be expanded.

Acknowledgments. This work was supported by Bialystok University of Technology under the grant S/WI/4/2008.

References

[1] Walsh, N.: DocBook 5: The definitive guide. O'Reilly, Sebastopol (2010)
[2] Goossens, M.: Writing documentation using DocBook: using DocBook at CERN (2000)
[3] Stayton, B.: DocBook XSL: The Complete Guide. Sagehill Enterprises (2007)
[4] Martínez-Ortiz, I., et al.: Using DocBook and XML technologies to create adaptive learning content in technical domains. Int. J. Computer Science Applications 3, 91–108 (2006)
[5] Draheim, D., Neun, W., Suliman, D.: Classifying Differential Equations on the Web. In: Asperti, A., Bancerek, G., Trybulec, A. (eds.) MKM 2004. LNCS, vol. 3119, pp. 104–115. Springer, Heidelberg (2004)
[6] Bzyl, W., Przechlewski, T.: Migrating to XML: The Case of the GUST Bulletin Archive. In: Syropoulos, A., Berry, K., Haralambous, Y., Hughes, B., Peter, S., Plaice, J. (eds.) TUG 2004. LNCS, vol. 3130, pp. 169–178. Springer, Heidelberg (2004)

Augmented Reality as a Space for Presenting and Passing the Information about Works of Art and Architectural Monuments

Jarosław Andrzejczak and Rafał Szrajber

Abstract. Fast-developing multimedia technology (especially in the sphere of Internet publication) allow us to cross borders associated with popularizing the works of art as well as architectural monuments. Once famous masterpieces could first and foremost be found on pictures and films. Later the virtual worlds appeared, in which one could find and explore virtual museums together with its exhibits. At present we may just find ourselves in a place where e.g. historic building used to stand in see it using the augmented reality. Not only can the augmented reality expand the scope and method of reception of information but also enlarges the number of its recipients due to the development of technology supporting it. Therefore, it becomes a medium of a huge scope of influence. The article constitutes the ideological basis for beginning the research on the use of augmented reality in the presentation and passing the information about works of art and architectural monuments.

1 Augmenting the Reality

Augmented Reality (AR) constitutes a part of the virtual continuum, whose model of continuity progresses from the Real Environment (RE) to the borders of the Virtual Environment (VE). The space between those environments is called a sphere of Mixed Reality (MR). Contrary to the Virtual Reality (VR) this extension does not create a new, fully virtual 3D world, but complements the one we know with the virtual content [1].

Hence this completion of the real world turns out to be a substance, which can take many different forms – from the simple information to complicated photo-realistic an interactive models or animations. Similarly to the virtual reality and developing technologies linked with it, the "augmentation of reality" leads to making the presented content "real" and connecting it with the surrounding real environment. In such a case we get the representation of the real environment by projecting the picture obtained by means of a camera, which is supplemented in

Jarosław Andrzejczak · Rafał Szrajber
Institute of Information Technology, Technical University of Lodz
Wolczanska 215, 90-924 Lodz, Poland
e-mail: andrzejczak@ics.p.lodz.pl, rafals@ics.p.lodz.pl

P. Lipiński, K. Świrski (Eds.): Towards Modern Collaborative Knowl. Sharing Syst., SCI 401, pp. 49–60.
springerlink.com

real-time by virtual elements containing additional information about the physical object [2]. For instance, when we are next to an architectural monument and we have appropriate application installed in our mobile phone, we can simply direct the mobile camera towards the monument and the picture of it along with additional information provided by the application will appear on our phone screen. The earlier mentioned connection happens in many different levels, from the simple recognition of markers by enhancing the traditional printed materials, to movement detection systems or links with the complicated technological structure used in medicine or military. It is also essential to pay attention to the development of augmented reality as a space for passing and supplementing the information in relation to mobility (which is a term used to describe a process of translocation as an element of culture as well as the phenomenon which accompanies the development of modern technology[3]) in which AR plays a more and more vital part as the navigation equipment uses the AR technology more and more frequently.

2 Digital Information about the Work of Art or a Monument in an Augmented Reality

Digital revolution, which enabled easy and widespread access to the Internet, also extorted the change in the approach towards the appropriate documentation of the works of art and architectural monuments [4]. Presenting the works of art in a form of digital information is more and more often transferred beyond the frames of the screen, which up till now used to be the basic level of communication with the virtual world. The fact of "separation" from the traditional computer screen allows for an absolutely different contact with the received information as well as broadens the offer of galleries and museums. This significant change however requires the tools and technology, which would allow for the connection of both worlds. Larger access to cameras (built in popular devices as well as the external ones – mainly the Internet cameras) as well as monitors (including the touchscreen ones) linked to them drew the augmented reality closer to the statistical computer user and made its use fuller.

Passing and popularization of the information about the works of art by means of AR systems occurs in many levels (similarly to the presentation of the works of art on the virtual space [5]):

- **direct** – which by the visual information constitutes the exact copy of the work of art together with showing its mobility, which is a result of the receiver's actions and the environmental influence. This level allows for an illusory contact with the presented object.
- **indirect** – which makes it possible to present the interpretation of the work of art, to ascribe characteristic features to it (e.g. for search engines). Moreover it also gives the rules of presentation and order of its elements (e.g. the change in displaying the details in the course of exploration). This level constitutes the educational and descriptive value, which provides knowledge about the presented object.
- **web** – which functions through pointing towards the connections of the digital content with other information (e.g. about the author, style the work of art represents)

and in the case of e.g. the lost works - it points to the sources on which the work was reconstructed.

- hidden – the level created by the user as a record of his/her relations with the presented object (e.g. the record of movement and placement of the work of art, a personal digital map of interaction).

The above mentioned description of requirements shows that the preparation of the work of art to put it in the augmented reality requires a huge workload. The analysis of the digital information as well as its processing not only contains the aspect of the work of art's projection, but also an info-graphic presentation of the whole spectrum of its features and connections. The augmentation of space allowed for supplementing it by diverse information, which, together with the development of technology, are much easier to pass and make interactive.

Augmented reality allows for a different level of association with the work of art. It helps to take the exhibit from its showcase and give it to the hands of the user. It enables interaction with nit as well as presentation of the huge installations in a smaller scale. In the open space the augmentation of reality allows for supplementation of the vision of the world around us with additional information projected by e.g. linking it with the GPS systems.

Yet, it is necessary to remember, that such a usage of augmented reality is not a way to save that works of art or monuments, but only a form of documentation (collection of digital data) and presentation of the received outcomes (the analysis of the digital data and the way of presenting it stands for a digital filter).

The research we are conducting is aimed to set the requirements for the application using the augmented reality, which would allow for proper "link" to the digital information about a work of art or a monument. Describing the appropriate rules essential for collecting and popularizing the information linked to the digital model of the work of art or a monument will increase its attractiveness and its reception. It will also improve the way of gaining knowledge about it. This way of organizing the presentation will allow the recipient to interact with objects of huge artistic and material value on his or her own, which is very difficult if not impossible n a real museum. In the same time it might increase the possibilities of the e-learning systems by advanced interaction by means of standard computer equipment.

3 Chosen Applications of Augmented Reality in Connection to Available Technology

Collecting the data about the work of art or a monument usually occurs before the direct contact with it. Very often however it turns out that after seeing the work of art (or sometimes during the observation) the need to supplement the knowledge of its internal structure, iconographic meaning, aesthetical value and historical context appears.

One of the possible uses of augmented reality is replacing the multimedia publications added to books in a CD or DVD format with AR technology. In such a case instead of browsing though the CD (or DVD) file tree or going through the menu of the application we may simply direct the book page towards the camera (or the other way round) and see the object in 3D together with an additional information on our computer screen. Since such innovation requires the use of appropriate tool or application, which would catch the picture from the camera and project the proper object on the screen, the final outcome of such an improvement may seem too small comparing to e.g. the use of CD, DVD or any other carrier of information. However it is enough for us to make the above mentioned application accessible on our publication or book website and we will put its address next to the AR marker. Then the user enters an appropriate website and does not need to install any additional software. A good example of such solution is the book presented in the augmented reality [6].

In the case of augmented reality with the use of camera and the Web the question of compatibility with different browsers arises as well as the accessibility of the service. Such a service however is very often solved by the Flash technology. Application prepared in such a way will fulfill all the functional conditions and also (due to the fact that Flash operates on vector objects) will be accessible to the majority of internet users (according to the research and statistics presented in [7]). Additionally it will enable the interaction with the projected objects, which for obvious reasons is impossible in the case of static pictures (both printed and projected on the screen). It is easy to imagine a regular art album, which enhanced by the augmented reality in the form of application and marker printed on its specific pages, will create a new quality – an interactive album (similar to the book discussed earlier). Therefore thanks to the change of the point of observation connected to the turning and moving the album itself, the 3D sculptures will be observed. Furthermore, the access to additional content about such objects will enrich such presentation and allow for plunging into the world of the unknown and new links.

Mobile devices especially mobile phones [8], [9] are another interface with a huge potential to use together with the AR technology. At present more and more users have mobile phones, smartphones or tablets with big screens that fulfill other system requirements, which enable the use of the assets of the augmented reality. For instance the mobile devices can be used to present the objects in their proper place, before they are even realized or do not exist in that particular location anymore (in both cases we only have a digital model of the information about them).

A similar idea would be visualizing the furniture or decoration objects for the customer in real-time in his own home [10], [11]. It increases the possibilities of e-business in the case of companies from the furniture, architecture or decoration branch. Such an idea may be used while designing and equipping the virtual museums and galleries directly by the users. Newly-emerging tools and solutions for mobile platforms seem to support the fact that the trend of using augmented reality is legitimate and perceived as a development. Google Android system and the ASR SDK [12], [13], [14] created for it serve as a good support for that thesis.

"Domesticating" the virtual reality allowed for its rapid development. The observation of the development of technologies, especially the mobile ones, lets us foresee a similar future for the AR systems as well.

What is crucial to mention is the fact that such a software was introduced in the National Museum in Cracow, where the application AR mGuide has become a mobile 3D guide in a mobile phone. Thanks to the application the pictures, videos and other multimedia content is presented on the basis of the markers recognized by a camera and which are found in the telephone. It is a very interesting example of changes that occur in the presentation and accessibility of the works of art by means of devices and media known to the typical user. Technology and mobile devices' accessibility is growing day by day which makes this market a priority for every action connected to processing and presentation of information.

Augmented reality is a totally new space for information visualization. While mobile devices do not allow for photo-realistic presentations of models as yet, AR makes it possible to bring the information content in real-time over the picture of a physical object (recorded by means of the camera – as it was described in the example in the first chapter). The observer has the impression of having the information directly over the physical object. Such approach significantly accelerates the understanding of the message by the recipient as the process of object transformation from the virtual system to the real one is eliminated. What is more, the 'clarity' of presentation improves, because there is no need to put the information content in a way that would disturb the vision of the presented object (e.g. information being too close or covering the object).

Augmented reality is probably a part of the digital revolution, which constitutes a long-term evolutionary process of technological and media discoveries.

4 The Disadvantages of Traditional Ways of Presenting Works of Art and Monuments

The greatest problem with the presentation of the works of art is a very frequent lack of interaction possibilities with it. Many works (especially from contemporary art) reveal their inner structure and artictic value when a person observing it changes from a passive observer into an active participant. Unfortunately the risk of destroying and damaging the work is too high for galleries or museums to decide on full interaction of viewers with the objects they own.

Such behavior of these institutions is understandable from a certain point of view, but in the same time it lowers the cognitive and aesthetic value of their exhibits. We called this conflict the problem of "the exhibit behind the glass" or "the exhibit in a show-case".

The solution of using the augmented reality to pass the information about the works of art and monuments, which we suggested, provides an answer to the problem by replacing the real object (the original) with the virtual one (a digital copy) and enabling the recipient a proper interaction with the given object. We have to remember however that the true localization of the object is retained thanks to the picture of the surrounding taken from the camera. It was proven earlier that this fact distinguishes the augmented reality used in this case from the virtual reality.

The next issue, which occurs while observing the works of art and especially monuments of a complicated structure I the lack of information balance[1].

[1] Information balance is understood as a state in which the amount of information getting to an addressee is regarded as sufficient by him/her (guarantees full comprehension). The disturbance of this balance is achieved through an excess of content or its deficiency. This state is not easily achieved with standard forms of presentation – it is much easier to be achieved using interactive systems.

While watching a monument we usually want to get to know as much as we can about it, but we also require the information to be given in an interesting, straightforward and easy way. Thus the description of an object and its crucial elements should not be presented e.g. in a form of long blocks of text on the information table. Such a way of placing the information will probably be treated as distracting our attention from the described object. On the other hand though the lack of direct relation between the information content and an object may result in lesser understanding and difficulty with the reception of the object (e.g. if all the elements of the interior are described on one information table, we have to be very careful when we reach a specific fragment of this interior because we may have a problem to distinguish which fragment of text referred to it).

Ambiguous message appears also when too much information is given in one place or when the information is illegible. Additionally, in such a case the recipient focuses mainly on trying to understand what the information description refers to instead of fully enjoying the contact with the monument.

Such a situation happens when for example we gather information from the leaflet or a catalogue. In such papers one can only find a two-dimensional scheme of the object, which is usually its projection or section. Then, the conversion of the flat plan into the real-space or imagining the 3D space on the plane is necessary.

Next, the process has to be repeated with every change of the place of observation. The described problem may be eliminated by means of augmented reality. The recipient gets a digital model of e.g. interior, art object, monument etc. with the basic information. In the same time this model contains interactive elements allowing for getting detailed information. Those elements are for instance special, visible markers which have meaningful and contrasting form (to be noticed easily by the viewer). Among the other types of elements performing this function, one can find invisible markers (i.e. masks - allowing for interaction when in contact with a specific fragment of the model), the model itself and its appropriately presented elements.

We focused on the issue of ensuring the interaction in a more detailed way in the sixth part of the article.

Information material do not have to be limited only to the text form – the picture form is also possible (e.g. the archive photos concerning the view of the original object and recording the further renovations) as well as drawings (e.g. schemes, architectural plans), videos (recordings presenting the building or renovation process etc.) or even audio recordings (with a commentary supplementing the description or the history of an object).

5 Augmented Reality Rich with Information

It was not without a certain purpose that we previously spent a lot of time on explaining the issue of putting and passing the information about the works of art and monuments. This piece of information is a key to create an illusion of total connection with these objects. A connection of this type should be focused not only on visual impressions, but also on cognitive values and better understanding of the new sphere of knowledge [15]. Therefore the use of augmented reality cannot come down only to the static presentation of the digital object, as only during the interaction with this object and during the assimilation of the information about it is the new quality created. New patterns and connections arise, which up to this point we could not notice and which we could not experience while familiarizing with the work of art in a different, superficial way.

As a support for this thesis we can refer you to the article [15], which focuses on a dynamic architectural representation of Cracow's heritage, and also the article [16] from 2003. In the latter the authors suggest the term of "IRVE" – Information-Rich Virtual Environments. The use of this term in relation to the virtual environment only makes sense when it contains information which are accessible perceptively (by means of senses e.g. the color of objects, the size) and also the abstract content with a huge information load and which are impossible to gather only by means of senses (e.g. the history of the object, technical schemes etc.)

The system or the presentation which fulfills the above mentioned criteria and is free from the described mistakes we called Information-Rich Augmented Reality (IRAR for short), which is similar to the terminology used in [16].

Hence such a system should:

- ensure the information balance
- deepen the impression of contact with the work of art or a monument
- make as much information accessible about the work of art or a monument as possible by means of many different media
- ensure high level of interaction together with keeping the service intuitive
- use the abstract information (according to the definition in [16][2])

6 The Ways of Presenting Detailed Information about the Object on the Example of Chosen Technologies

In part 4 of the article we gave three ways of putting the information on a virtual object by means of interactive:

1) markers in the form of objects
2) invisible markers / masks
3) the model itself or its part

In the first case small, contrasting objects of characteristic shape are added to the model. The size and shape of these objects must ensure its identification in the virtual environment or film space (in the case when we augment the picture collected from the camera – Fig.1) as separate elements, which are not parts of the model itself (Fig.2 and Fig.3). This solution is very comfortable while creating a model, as it

[2] Abstract information is understood as a one which is not accessible perceptively in a physical world (e.g. the history of an object, its creator, technical data of the material etc.).

does not require any intrusion from the author in the structure of the model itself. It also facilitates the reception for the viewer, because the person is informed in a clear way of the amount of accessible detailed content, its location in the environment and its connection to the model. Yet, in some complicated pictures or scenes the problem of "buzzing" and excess (redundancy) may appear which in turn leads to the problem of distortion of information balance as a result of using too many markers. Thus, such solution should be used when the amount of detailed material that is presented to the viewer is not too vast.

Fig. 1 Marker in the form of an object used in picture collected from the camera

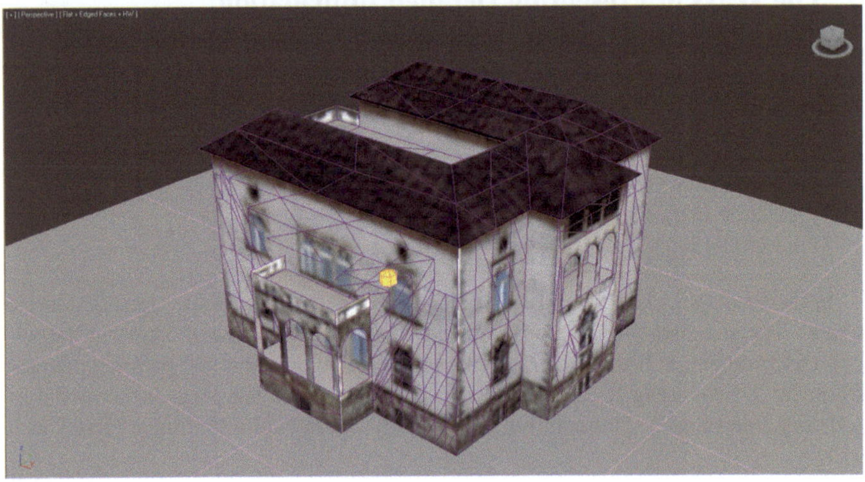

Fig. 2 Marker in the form of an object during creating (Autodesk 3DS MAX 2011)

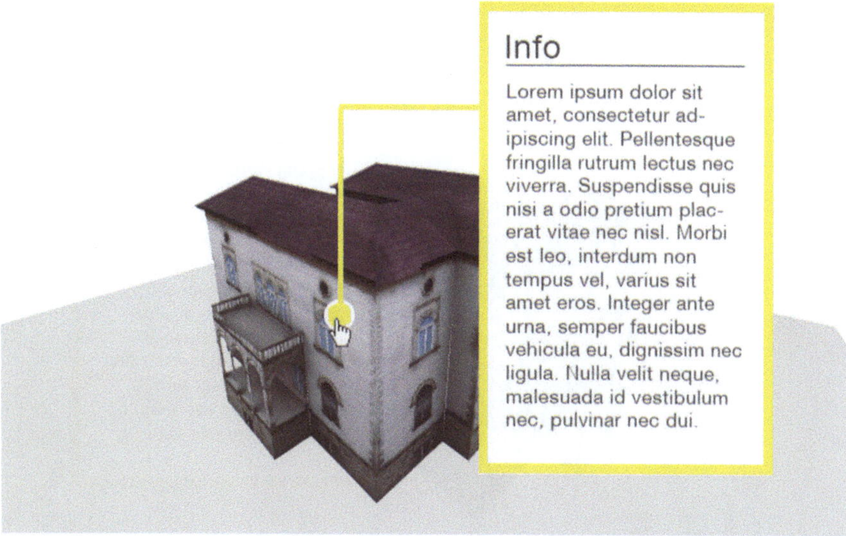

Fig. 3 Marker in the form of an object used in interactive application (Flash Player 10.1)

The second method uses the invisible markers, which could be understood as "masks" fixed directly next to the appropriate fragment of the model (Fig. 4 and Fig. 5). This solution is somehow the opposite of the previous one. In this case the problem of distorting the message by too much information does not exist as the markers are transparent and thanks to that they do not cover the model. However in this particular situation another problem arises i.e. how to suggest the recipient where to look for additional information (which of the elements can he/she interact with?). A hybrid solution would be efficient here i.e. linking this method with the previously mentioned one, that is to use e.g. translucent objects or objects with frames (Fig. 6 and Fig. 7).

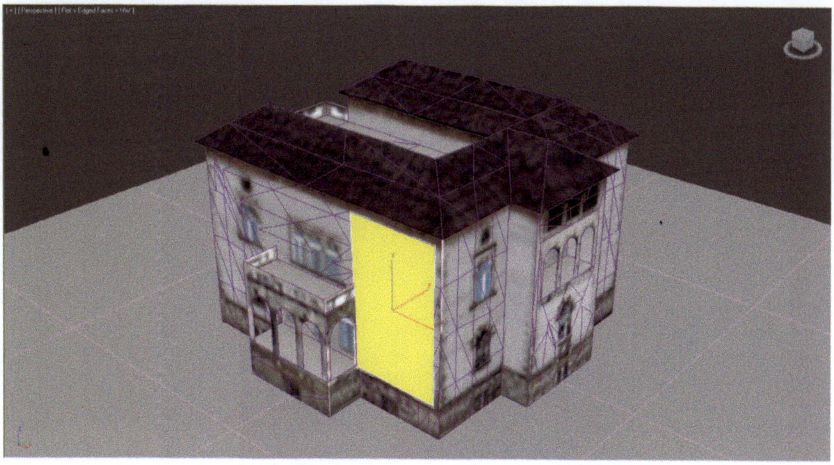

Fig. 4 Invisible marker/mask during creating (Autodesk 3DS MAX 2011)

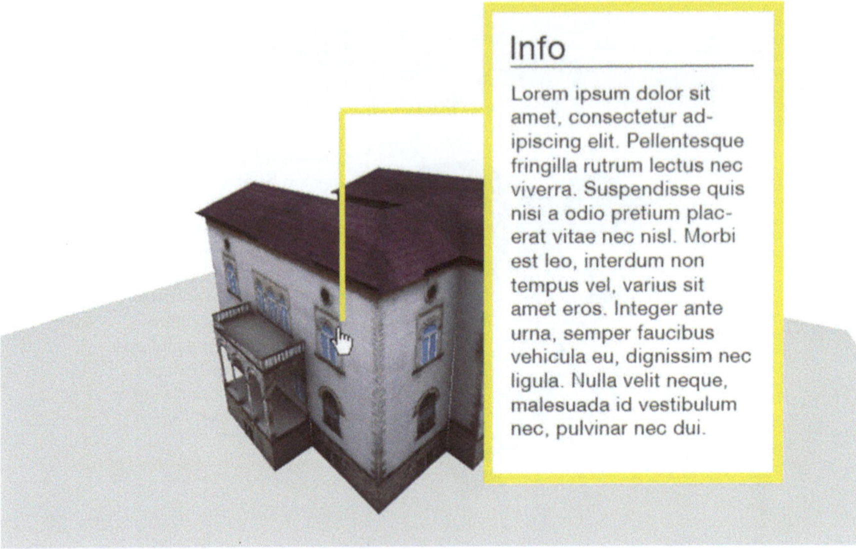

Fig. 5 Invisible marker - mask used in interactive application (Flash Player 10.1)

The third method is based on converting fragments of the model into interactive elements (Fig. 8). It is a method which requires the largest amount of work from the author of the presentation, as the model needs special "cutting" and separation of the interactive parts. It is also extremely hard to implement it into a given application as two spheres are mixed: the sphere of the presentation of the object and the sphere of detailed information about it.

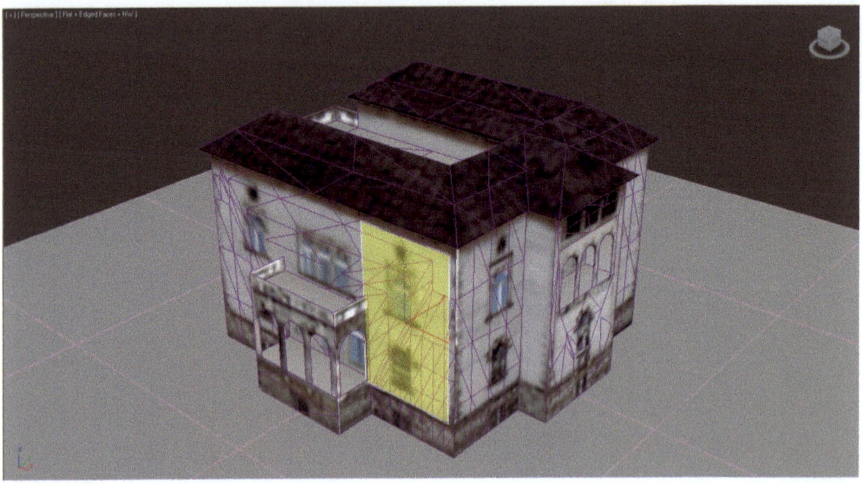

Fig. 6 Hybrid marker during creating (Autodesk 3DS MAX 2011)

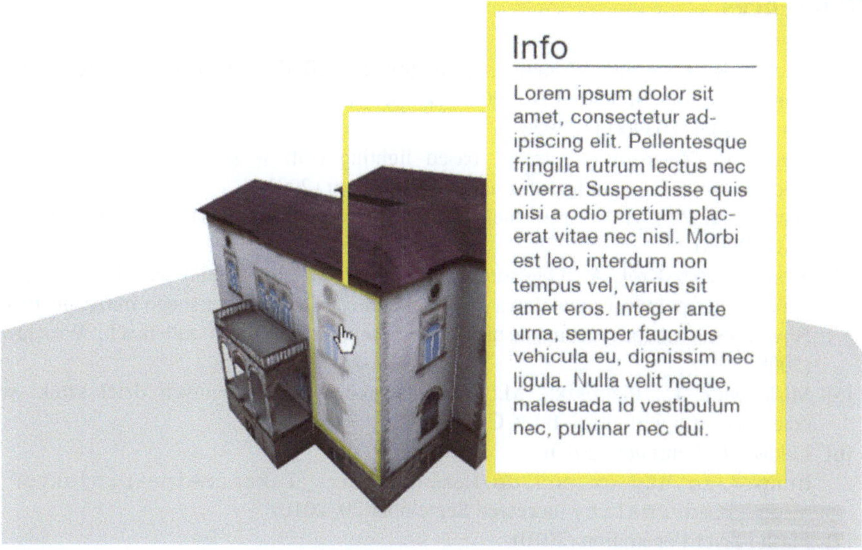

Fig. 7 Hybrid marker used in interactive application (Flash Player 10.1)

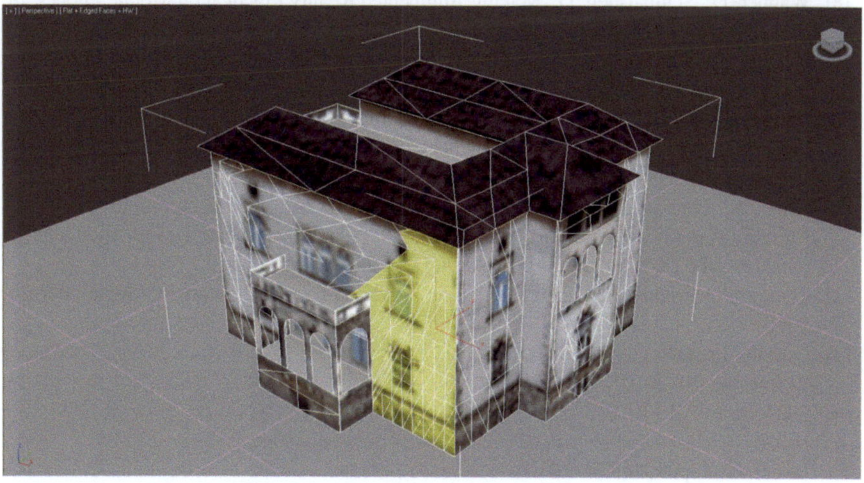

Fig. 8 Model components as a marker during creating (Autodesk 3DS MAX 2011)

Summing up, we have to point out that the first two methods can be used to fulfill the criteria of IRAR, both in case of placing the content over the virtual object and while adding the information to the real-time scenes (the picture taken from the camera supplemented by the content on the display of the device). Even though the third method seems to be the most intuitive for the recipient, it is extremely hard to realize for the author, requires vast amount of work and is very difficult to use in case of supplementing the picture from the camera by the information content (that is when we do not use the digital model).

References

[1] Azuma, R.: Overview of augmented reality. In: GRAPH 2004, Proceedings of the Conference on SIGGRAPH 2004, Course Notes, New York, vol. 26 (2004), doi:10.1145/1103900.1103926

[2] Pardel, P.: Zastosowanie image based lighting (ibl) w środowisku rozszerzonej rzeczywistości (ar). Studia Informatica (4), Gliwice (2009)

[3] Gawrysiak, P.: Cyfrowa rewolucja. Rozwój cywilizacji informacyjnej. PWN Warszawa (2008)

[4] Kusmidrowicz-Król, A.: Odwzorowanie cyfrowe dzieła sztuki i techniki multimedialne – perspektywa rozwojowa czy konkurencja dla współczesnego muzealnictwa. Nowoczesne metody gromadzenia i udostępniania wiedzy o zabytkach. Wrocław (2008)

[5] Miskiewicz, L., Pietruszka, M.: Prezentacje czasoprzestrzennych dzieł sztuki w rzeczywistości wirtualnej. In: ICCVG (2002)

[6] Lukas, M.: Lehrbuch (2010), `http://ig.hfg-gmuend.de/Members/mark_lukas/meine-projekte/augmented-reality` (accessed September 20, 2010)

[7] Flash Player Penetration (2010), `http://www.adobe.com/products/player_census/flashplayer/` (accessed September 21, 2010)

[8] Wagner, D., Schmalstieg, D.: Making Augmented Reality Practical on Mobile Phones. In: Part 1. Computer Graphics and Applications. IEEE (2009), doi:10.1109/MCG.2009.46

[9] Wagner, D., Schmalstieg, D.: Making Augmented Reality Practical on Mobile Phones. In: Part 2. Computer Graphics and Applications. IEEE (2009), doi:10.1109/MCG.2009.67

[10] Wang, R., Wang, X.: Applying Service-Oriented Architecture into an Augmented Reality E-business System. e-Business Engineering. IEEE (2008), doi:10.1109/ICEBE.2008.103

[11] Chen, A., Huang, Y.-L., Chen, C.-H., Tai, K.-C.: Visual augmented reality interface for table-top e-business environment. Computer-Aided Industrial Design and Conceptual Design (2006), doi:10.1109/CAIDCD, 329463

[12] Google Android (2010), `http://www.android.com/` (accessed September 20, 2010)

[13] Google Projects for Android (2010), `http://code.google.com/` (accessed September 20, 2010)

[14] Qualcomm AR (2010), `http://developer.qualcomm.com/dev/augmented-reality` (accessed September 20, 2010)

[15] Blaise, J.-Y., Dudek, I.: Dynamic representations for information visualization and retrieval: a case study on Krakow' s architectural heritage. In: 26th International Conference on Information Technology Interfaces (2004), doi:10.1109/ITI.2004.241873

[16] Bowman, D.A., North, C., Chen, J., Polys, N.F., Pyla, P.S., Yilmaz, U.: Information-rich virtual environments - theory, tools, and research agenda. In: Proceedings of the ACM symposium on Virtual Reality Software and Technology (2003), doi:10.1145/1008653.1008669

3D Scene Exploration with External Controllers

Adam Wojciechowski

Abstract. For some users human-computer interaction is identified with mouse and keyboard usage. However interaction within three dimensional environments can be also performed by means of external input and output controllers. In spite of professional devices dedicated for navigation and interaction in virtual environments there is a relatively wide range of cheap and available controllers inherently used in computer or video games. Besides that their functionality and precision is constrained, due to adequate software methods they can be used for advanced exploration tasks. This paper aims to present selected interaction methods and their implementations.

1 Introduction

Since the WIMP (Windows, Icons, Menu, Pointer) interface introduction noth-ing seemingly has changed in the human-computer interaction field. Just seem-ingly, as computers' users consciousness has matured to exploration of artificial three dimensional environments and even demands three dimensional experiences. It might be easily noticed, starting from the content of contemporary web pages analysis (spatial layouts), through three dimensional movies and films and on computer games finishing. Though main technical barriers of a three dimensional visualization has already been broken the problem of easy and intuitive spatial interaction still remains. The problem probably originates from the fact that one universal mode of interaction for all types of spatial scenes and objects has not been invented yet and probably does not exist. Different tasks within virtual environments require individual approach and interaction devices provide miscellaneous and very often restricted functionality. Moreover human-computer interaction tools are not standardized and researchers still look for a most convenient hardware and software composition.

Presented paper suggests selected methods of specified tasks completion by means of external controllers. Although most of controllers are very expensive and though unavailable, there is still limited number of devices quite available for casual users. Their costs should not exceed one hundred US dollars. Even though they have

Adam Wojciechowski
Institute of Information Technology, Technical University of Lodz
Wolczanska 215, 90-924 Lodz, Poland
e-mail: adam.wojciechowski@p.lodz.pl

P. Lipiński, K. Świrski (Eds.): Towards Modern Collaborative Knowl. Sharing Syst., SCI 401, pp. 61–70.
springerlink.com

evident drawbacks - their functionality and precision is limited and access to appropriate programming API is restricted - fortunately interaction process, utilizing such equipment, can be improved both by smart task-device association or innovative methods of interaction that support user while interaction.

2 State of Art

Human-computer interaction innovative solutions were initially elaborated in virtual environment laboratories. Some people regard Ivan Sutherland as a father of virtual environment. He was an author of Sketchpad [1] system, where owing to invented head mounted display users were able to explore three dimensional artificial environments. In 1980 it was introduced a voice command steered interface called Put That There [2]. Five years later two systems VideoDesk and VideoPlace, based on two dimensional hands movement tracking, were presented. Hands were used for pointing, drawing or dragging objects. In 1985 Lanier [3] has developed system of goggles and data gloves regarded as achievement of the decade. Year 1995 has abounded in a three dimensional electromagnetic tracking systems, that allowed to track a hand or a whole body position and orientation without additional controllers [4].

All of mentioned examples represented quite expensive and rather unavailable solutions, especially for ordinary computer users. Such situation took place until the beginning of the XXI century. Constant development of a computational power and progressing miniaturization brought an alternative mean of interaction especially in a video games field [5,6,7]. Those solutions were quite hermetic due to their usage limitation only for a specific environment of games' producers.

Nevertheless growing popularity of computer games and constant interest in associated technologies development caused remarkable growth in miscellaneous controllers' availability. Permanent interest of researchers and academic groups provided tools and libraries for external controllers programming. However there is still a noticeable deficiency of intuitive methods for experimental and innovative task completion. That is why one of the goals of this paper is a selected tools study and equipment functionality analysis and their usefulness for specified human-computer interaction task completion. As a practical goal of the paper a specialized set of tools and libraries must be mentioned. Presented solutions would be helpful for researchers tending to control their applications by means of alternative interfaces.

3 Method

Methods developed during the project would exploit external controllers providing simple motion tracking, accelerometers, gyroscope or computer vision systems. Exemplary controllers: Sony Dualshock 3 (fig. 1a) and Sony PlayStation Eye (fig. 1b) are presented in a figure 1.

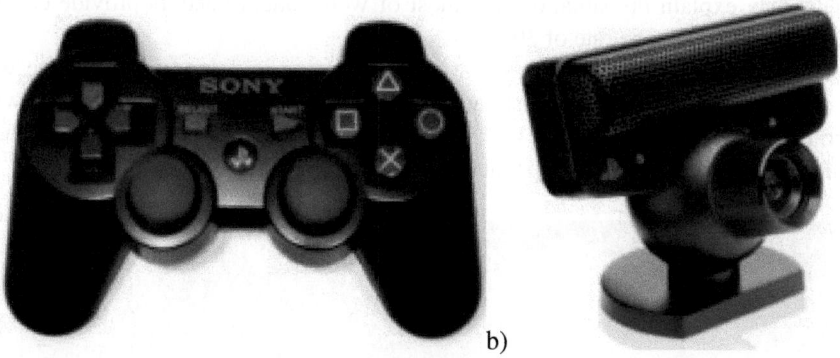

a) b)

Fig. 1 Exemplary cheap controllers: a) Sony Dualshock 3, b) Sony PlayStation Eye.

One of the project's assumptions is to minimize interface costs, by making use of mainly simple and cheap hardware controllers, while maximizing user performance, system usability and reliability. Selected controllers are programmed with dedicated API or additional specialized libraries: i.e.: OpenGL, HID, CL-eye, OpenCV. Possibly simple physical engine (SPE) was used for extending application functionality.

Several subsets of controllers was used for controlling several minor applications, in which user was able to complete logically–skilful tasks like: playing tennis, fighting with a sword, brush painting, puppet animation control or objects' interaction. For each application suitable interface and interaction mode was designed and implemented. All devised methods and designed concepts were implemented and their usability was tested.

Hardware implementation required usually additional signal denoising process and adequate controller mapping into controlled object position and orientation.

By example Sony DualShock 3 controller records both accelerometer translation signal and gyroscope rotation signal with the same variable so parallel translation and rotation transmission is not possible. Additional destructive feature was noticed while controller resting. Besides its stable position (it laid on a table) it has generated constant signal noise what might resulted in controlled objects oscillations. Its working frequency (100 Hz) was sufficient for correction filtering and streamlining movement fluency. Despite the fact that controller has in its name six axis lack of one degree of freedom was noticed. It does not provide direct control over its vertical axis (Y) but its indirect approximation evokes many problems and instability. Moreover lack of any reference point while interaction caused constant recalibration necessity.

As a visual hardware controller Sony PlayStation Eye video camera was chosen. Its frame rate can reach 120 Hz for images 320x240 pixels in size, or 60 Hz for images 640x480 pixels in size. It supports bulk transfer mode with USB 2.0 hardware interface, what cause transfer bandwidth problems. The size of 60 true color images considerably exceeds USB 2.0 practical transfer mode (40 MB/s).

This may explain the situation that most of web camera vendors provide camera with maximum frame rate of 30 Hz.

For visual tracking systems suitable color coding as a key tracked value was exploited. Color markers were made individually basing on a torch ended with color semitransparent plastic ball (fig. 2). Such marker construction guarantees different color usage for independent tasks.

Fig. 2 Colored markers made of torch ended with semitransparent plastic ball.

As a source of motion tracking methods OpenCV library was used. It provides wide range of feature tracking algorithms. Starting from Haarlike features based face detection and face tracking Viola-Jones method [12], through motion templates, Kalman motion estimation algorithm [13], Lucas-Kanade optical flow method [14], point of interest detection and description with SURF algorithm [15] and on, histogram based, Camshift algorithm [8] amidst others finishing. Due to colored balls usage the last method was mainly exploited, however its functional improvements were introduced.

First step lied in finding, within an image, selected color pixels and their possible similar neighbors. For this stage of the tracking algorithm, inherent OpenCV `cvFloodFill` method was used with three independent color value tolerances (for three selected colors).

```
cvFloodFill (image,
                cvPoint (fw ,fh ), cvScalar(255 * m ,0 ,255 ),
                cvScalar(   (unsigned int)TmRGBTolerance[2],
                        (unsigned int)TmRGBTolerance[1],
                        (unsigned int)TmRGBTolerance[0]),
                cvScalar(   (unsigned int)TmRGBTolerance[2],
                        (unsigned int)TmRGBTolerance[1],
                        (unsigned int)TmRGBTolerance[0]),
                &flood_info,CV_FLOODFILL_FIXED_RANGE);
if (flood_info.area > 5){
        flood_stop = true;
        break;
}
```

One of the advantages of `cvFloodFill` method was providing a programmer with modified image description. Respectively it comprises area of marked region, its coordinates or region bounding rectangle. This information helps exclude too small regions. In spite of above, `cvFloodFill` method has some disadvantages. As marker has different color shades in its central and border area, just a part of the marker was occasionally recognized (fig. 3).

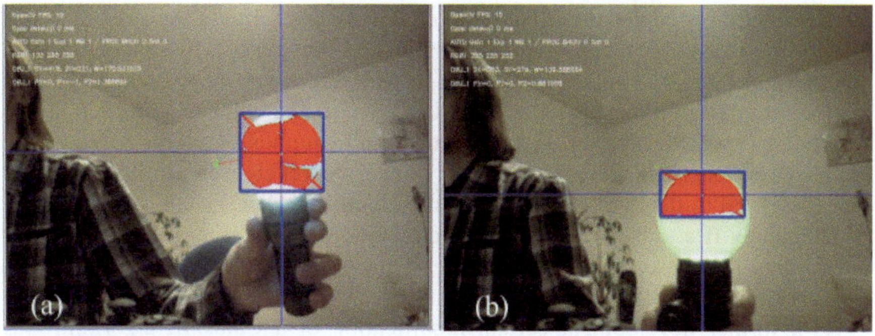

Fig. 3 Image transformed with `cvFloodFill` method. a) correct marker area recognition; b) wrong marker area recognition

Averaging signal has improved results but in a limited manner. As a method extension filling a region was exchanged for simple check whether specified pixel fall into color threshold. Method extension can be coded according to subsequent code:

```
int minx[2], miny[2], maxx[2], maxy[2];
for (fh = 0; fh < image->height; fh++)
   for (fw = 0; fw < image->width; fw++) {
   uchar* temp_ptr = &((uchar*)(image->imageData+ image->widthStep*fh))[fw*3];
      for (int m = 0; m < _iloscKO; m++){
          if (  temp_ptr[2] >= TmRGBColour[m][0] - TmRGBTolerance[m][0] &&
                 temp_ptr[2] <= TmRGBColour[m][0] + TmRGBTolerance[m][0] &&
                 temp_ptr[1] >= TmRGBColour[m][1] - TmRGBTolerance[m][1] &&
                 temp_ptr[1] <= TmRGBColour[m][1] + TmRGBTolerance[m][1] &&
                 temp_ptr[0] >= TmRGBColour[m][2] - TmRGBTolerance[m][2] &&
                 temp_ptr[0] <= TmRGBColour[m][2] + TmRGBTolerance[m][2] )
          {
                 if (fw > maxx[m]) maxx[m] = fw;
                 if (fh > maxy[m]) maxy[m] = fh;
                 if (fw < minx[m]) minx[m] = fw;
                 if (fh < miny[m]) miny[m] = fh;
                 break;
          }
      }
   }
```

where `image` stores current frame, and `temp_ptr` is a pointer to a currently checked pixel. As can be noticed algorithm saves independently the lowest (`minx`, `miny`) and the highest (`maxx`, `maxy`) pixel coordinates.

Algorithm can also track several markers made of different colors with separate color threshold. All variables are saved in two dimensional arrays `TmRGBColour[m][0]`, where `m` is a number of tracked color. A loop repeats only part concerning pixels' value so it is performed only once. Each pixel can be discarded or accepted and saved into one of the array fields. As a result pixels falling into two color ranges are assigned just to one marker. Advantage of such solution lets recognize marker even if colored torch shines directly into the camera (central part of a marker is perceived as a white and color is preserved just in external part of the marker shape). Even small number of pixels of specified color can reconstruct marker within an image, however even single accidental pixel assigned to the marker can drastically change marker's area.

Eventually if a scene was decently illuminated or markers were sources of colored lights a proposed solution provided satisfactory effects with recognition time not exceeding 10 milliseconds.

As a further development certain modification of a Camshift algorithm was proposed. It lets reduce noise while marker area reconstruction.

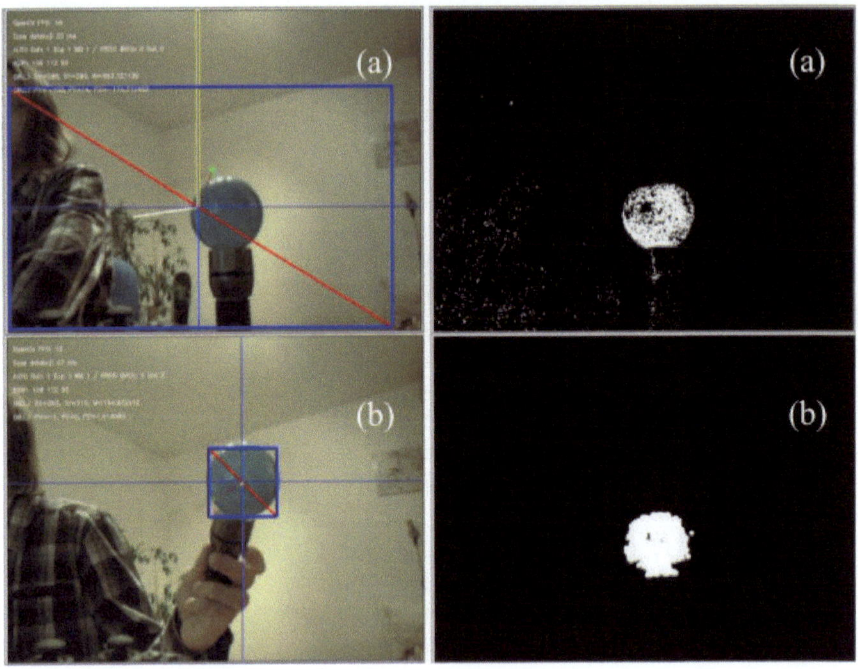

Fig. 4 Camshift based algorithm in action; a) image and its mask without denosing; b) image and mask with denoising

Camshift operates in HSV color space (`cvCvtColor` OpenCV method) and generates a mask (fig. 4) of a detected object (`cvScalar` OpenCV method) that can be operated independently from the image. This stage of the method was implemented before pixels classification. As the initial mask has very high noise rate (fig. 4a) it can be eliminated with two morphological operations: diliation (`cvDiliate` OpenCV method) and erosion (`cvErode` OpenCV method) (fig. 4b). So prepared mask became a source image for further pixel classification.

It must be remarked that third, Camshift based, modification method is relatively time consuming and considerably reduces application frame rate. Time necessary for a single frame processing increased up to 100 milliseconds, but its usage remained optional and it can be switched on just for dimly lit scenes with poor marker contrast.

4 Implementation

In this paper selected number of interactive application was implemented. First one concerned animation of a simple three dimensional puppet (fig. 5). Puppet mesh was assigned to the hierarchical spine made of bones, which were controlled with a colored marker. Marker translations cause corresponding bones' bending what simulates puppet's head movements. Animation of mouth, eyes, eyelids and eyebrows was predefined and stimulated with programmable Sony DualShock buttons. Pressing adequate button has evoked mimical facial movements.

Fig. 5 Animated puppet; a) bent to the left; b) unbent; c) bent to the right with their eyes and eye lids cast down.

Successive implementation has joined controller's features and physical engine functionality. Owing to camera marker tracking, application calculated strength and direction of a sword movement (cut). If sword has touched any object within a scene, its orientation, controlled with a DualShock pad, is considered. If sufficient force is applied object is cut into two pieces (fig. 6).

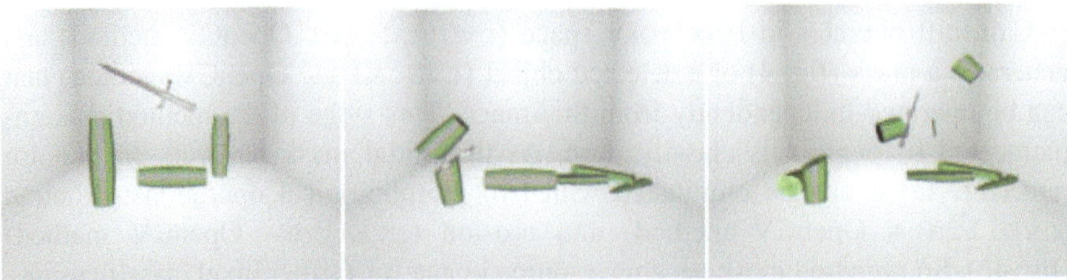

Fig. 6 Object disintegration made with a sword cut.

Objects' disintegration was supported with a Simple Physical Engine (SPE). If adequate force was applied objects were divided into several meshes and consequently simulated according to a gravity force.

Whatsoever different approach was applied in a third application. Driving a vehicle was conducted with pure hands tracking (fig. 7).

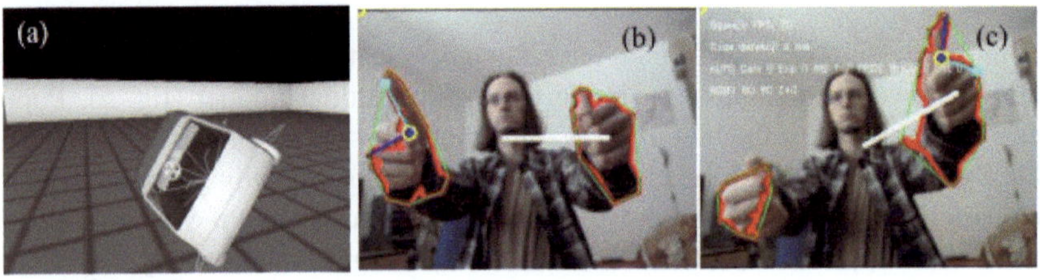

Fig. 7 Hand interface for a vehicle steering; a) car leaned over; b) hands' position causing car straight movement; c) hands' position causing turning left of the car.

Camera has registered hands' shape and their distance by means of image value channel in a HSV color space. For hands' segmentation, an adaptive background detection method [9] was used. Hands' movement was translated into vehicle movement (fig. 7a). Car direction and its turning range depended on the bending angel of the line crossing two hands contour (fig. 7b and 7c).

One of the most spectacular and interesting controllers' interpretation was interactive image painter (fig. 8).

Fig. 8 Drawing picture in a screen surface.

User could draw a picture with a visually tracked marker. Marker movements left adequately colored trail representing brush strokes. At the same time pad controlling 256 levels of pressure was responsible for RGB color space primary colors intensity determination. The distance of the marker form the camera affects its size and is interpreted as a brush pressure.

5 Tests

Besides technological problems, accompanying alternative human-computer interaction development process, users' point of view should be also taken into consideration. Researchers and developers of application interfaces should seek to improve modes of interaction from a user's perspective, ensuring their functionality and usability by following usability engineering methods. Before specific testing methods are presented and their results are given, it is important to set the context for usability testing. Developers of interactive applications often confuse the software engineering and usability engineering aspects. Well established validation and verification techniques should concentrate separately on software correctness and robustness and quite independent consideration of whether that software serves its users' needs. The quality of the user interface – the usability – of the interactive system is ensured by a user-centered focus on the look, feel, and behavior as a user interacts with an application [10, 16]. A quality of interface usability can be achieved by a user-centered development of user interaction components, comprising all icons, text, graphics, audio, video and devices through which a user communicates with an interactive application.

Usability testing typically consists of one or more usability evaluation or inspection methods. Usability inspection is a process that aims to identify usability problems in user interaction design, and hence, end use [11]. There are four generally accepted techniques for performing user task analysis: documentation review, questionnaire review, interviewing and observation [10]. Any combination of mentioned analysis techniques is sufficient for task improvement however adequate hard measurable tests data are strongly expected.

Hitherto tests of application usability were performed basing mainly on intuition and self-observation. Thorough analysis of user effectiveness and measurable rate of task completion satisfaction may be performed more extensively on a larger group of people. According to Nielsen [17], guru of usability testing, testing group should count between 5 and 9 people, what should eliminate more than 90% of usability faults. It lets improve controllers' usage for specified contexts.

6 Summary

Presented paper describes just a small range of ideas and methods elaborated during project realization. Besides a few abstract ideas most of presented solutions made human-application interaction more attractive. Devised methods were in fact designed and tested for specific context, however they are reusable and can be adapted to any invented interactive tasks. Each adaptation requires prior amendments and improvements before it is iteratively tested and approved. Second, worth mentioning, fact is that a total cost of exploited devices did not exceed one hundred US dollars at the moment of paper writing. It makes mentioned hardware available for most users.

References

[1] Sutherland, I.E.: A man-machine graphical communication system. In: AFIPS Conference Proceedings, pp. 323–328 (1963)

[2] Hinckley, K.: Haptic issues for Virtual Manipulation. Ph.D. thesis, Department of Computer Science, University of Virginia (1997)

[3] Gonazales, D.: Automation and Robotics for the Space Exploration Initiative: Results form Project Outreach, Rand Corp. Santa Monica CA (1991)

[4] Zimmerman, T., Smith, J.R., Paradiso, J., Allport, D., Gershenfeld, N.: Applying Electric Field Sensing to Human-Computer Interfaces. In: CHI 1995, pp. 280–287. ACM (1995)

[5] Sylvester, M.: Revolutionary Controller Showdown (October 10, 2010), `http://nintendo.joystiq.com`

[6] Lee, J., Hudson, S.: Foldable Interactive Displays. In: Proceedings of the ACM Symposium on User Interface Software and Technology (October 2008)

[7] Lee, J., Tan, D.: Using a Low-Cost Electroencephalograph for Task Classification in HCI Research. In: Proceedings of the ACM Symposium on User Interface Software and Technology (October 2006)

[8] Allen, J.G., Richard, Y., Xu, D., Jesse, S.J.: Object Tracking Using CamShift Algorithm and Multiple Quantized Feature Spaces. In: VIP 2005 (2005)

[9] Butler, D., Sridharan, S., Bove, V.M.: Real-time adaptive background segmentation. In: ICME 2003, pp. 341–344 (2003)

[10] Hix, D., Gabbard, J.L.: Usability engineering of Virtual Environments. In: Stanney, K.M. (ed.) Handbook of Virtual Environments, Design, Implementations and Applications, Stanney, LEA (2002)

[11] Nielsen, J., Mack, R.L.: Executive summary. In: Nielsen, J., Mack, R.L. (eds.) Usability Inspection Methods, pp. 1–23. John Wiley & Sons, New York (1994)

[12] Viola, P., Jones, M.: Robust Real-time Object Detection. In: Second Intl. Work-shop on Statistical and Computational Theories of Vision-Modeling, Learning, Computing and Sampling (2001)

[13] Li, Y., Ma, S., Lu, H.: Human posture recognition using multi-scale morphological method and Kalman motion estimation. In: 4th Intl Conference on Pattern Recognition, pp. 175–177 (1998)

[14] Lucas, B.D., Kanade, T.: An iterative image registration technique with an application to stereo vision. In: Proceedings of Imaging Understanding Workshop, pp. 121–130 (1981)

[15] Bay, H., Tuytelaars, T., Van Gool, L.: SURF: Speeded Up Robust Features. In: Leonardis, A., Bischof, H., Pinz, A. (eds.) ECCV 2006, Part I. LNCS, vol. 3951, pp. 404–417. Springer, Heidelberg (2006)

[16] Stobińska, M., Pietruszka, M.: The effect of contrasting selected graphical elements of a web page on information retrieval time. Journal of Applied Computer Science 17(2), 113–121 (2009)

[17] Nielsen, J.: Designing Web Usability: The Practice of Simplicity. New Riders Publishing, Thousand Oaks (1999)

System for Developing Photon Mapping Algorithms

Krzysztof Guzek and Piotr Napieralski

Abstract. This paper presents several algorithms for high-performance rendering engines. In the first part we describe the development of global illumination methods with particular emphasis on photon mapping. Secondly, we present methods used by the rendering system design and consider the possibility of further developing these methods.

1 Introduction

Rendering is a necessary stage in the process of any visualization. This is an important and multi-disciplinary issue, for commercial and academic work. The main task of our project is to create open source rendering software, for the purpose of speeding up and improving the rendering process.

Modern data visualization software is used both by the academic community as well as by commercial companies. Visualization services are used for architectural visualizations, as well as scientific and industrial processes in simulations. Advances in computer hardware and memory space have allowed the implementation of algorithms that enable the generation of high quality images with a high degree of realism. The most significant element of visualization is photorealism. It is very important in this type of process to simulate any optical phenomena while preserving the physical properties of light.

The basic assumption is the correct simulation of the way light rays travels from its source to the observer. Refraction and reflection of rays are calculated from basic optical laws and equations, respectively. It is necessary to solve the problem of following a single photon through its more than one scattering and/or more than one possible path from the source to the eye of the observer. However, every object will generate indirect illumination based on its actual material. Simulation of physical phenomena and properties of light in a synthetic model of

---

Krzysztof Guzek · Piotr Napieralski
Institute of Information Technology, Technical University of Lodz
Wolczańska 215, 90-924 Lodz, Poland
e-mail: krzysztof.guzek@p.lodz.pl, piotr.napieralski@p.lodz.pl

P. Lipiński, K. Świrski (Eds.): Towards Modern Collaborative Knowl. Sharing Syst., SCI 401, pp. 71–81.
springerlink.com

the world is called Global Illumination. This method allows finding the light intensity at any point in the space of a synthetic world.

Description of the 3-dimensional surface geometry is as important as the description of materials and light sources in a scene. Global Illumination describes the moment of leaving the source through a beam of light and its subsequent interaction with other objects of the scene. This algorithm allows the calculation of indirect illumination from the objects. This is the basic assumption of the lights phenomena simulation in the real world.

2 The Evolution of Global Illumination Algorithms

The development of global illumination algorithms going on over the past twenty years. The purpose of each method is to solve the rendering equation formulated in 1986 by James T. Kajiya [1]:

$$L(x',\omega') = E(x',\omega') + \int \rho_x(\omega,\omega')L(x,\omega)G(x,x')V(x,x')dA \quad (1)$$

where:

L(x',ω') – the radiation from the point x in direction ω

E(x',ω')- is emitted light from x' in direction ω'

ρx' (ω,ω') –BRDF function in the point x'

L(x,ω) – the radiation in the point x from direction ω

G(x,x') – geometric relations between x and x'

V(x,x') – visibility test between x and x'

Solving the rendering equation for any given scene is the primary challenge in realistic rendering. Along with discovering the equation the author proposed a technique to track light paths (called Path Tracing). This is an extension of the Ray Tracing method, which gives the possibility to calculate the total distribution of global illumination. The algorithm is based on Distribution Ray Tracing, which uses stochastic sampling to calculate effects such as soft shadows, motion blur, and depth of field. Path Tracing extends the idea of stochastic sampling of all possible paths of light, by averaging large numbers of samples for the pixel in the output beams. In this way we obtain an estimator of the integral given in equation (1). Path Tracing is a recursive algorithm, which has an exponential increase in the number of rays. This is in order to reduce the number of samples used to track only one radius because of the diffuse reflectance. This allows for calculation of reflections only from planes visible by an observer. This method is subject to variance estimators, which entails the creation of noise in the final image. Noise reduction is associated with a significant increase in computing time. The elimination of computing cost and the aforementioned noise are the main assumptions in global illumination research. Until now there have been many methods developed making the reduction of these problems possible.

A direct extension of Path Tracing is the Bi-Directional Path Tracing method. It consists of simultaneously sending the rays from the observer and the light source. In the final process the input of the vertices to the vertex path of light coming from the path of the observer is calculated. The method is useful for calculating global illumination for scenes with hidden light sources, also known as caustic effects. Similarly to the Path Tracing method, noise appears in the final image. It should be noted that rendering using the described method runs the risk of a large number of errors in the case of open scenes.

The application of different sampling methods allowed to significantly reduce the above inconvenience. The introduction of sampling, depending on the radiation has been used in the algorithm of Metropolis Light Tracing [2]. The use of such a solution allows for the automatic increase in the concentration of radiation in the lighter regions of the image. The algorithm starts from a random sampling over all light paths in the model (using the method of Path Tracing). These paths are then randomly cloned and modified (mutated). If the recently mutated path falls outside the scene, it is rejected and the current path is used again for a new mutation. If the new path is good it is then used as a new candidate for the base path. Calculated paths are used to calculate the final light intensity.

An alternative method to methods based on Path Tracing is Radiosity [2]. This solution uses the engineering field of heat transfer. Algorithms based on this method involve the behavior of light energy in a closed environment. All the energy is emitted or reflected from every surface, with assumption that it can be absorbed well by the other plane. In contrast to conventional methods of rendering, this methodology one must first determine all the interactions of light in the environment in a manner independent of the observer. Then it is necessary to execute the rendering to one or more relapses, which involves an additional calculation of area of visible and interpolation shading. The Radiosity method has been extended to the more complex reflection models. It should be noted that despite these efforts it has still not been possible to achieve the right results. One of the main problems is the length of calculations. To make the Radiosity technique less expensive, it was widened by View-dependent Calculations, clustering and hierarchical techniques. These techniques allowed to reduce the cost calculations, by removing redundant paths and planes that do not affect the final image. Radiosity is very effective for diffuse models, but it becomes costly and uneconomical for the specular surfaces. Expensive methods to eliminate the problems of this algorithm make it uneconomic for the more complex models. The goal of creating a hybrid technique was to connect all the described methods to maximize the benefits of some techniques and eliminate the weak points of others. Such methods are called multi-pass algorithms [2]. In this solution, Radiosity is merely the initial step of a progressive rendering process to calculate the indirect lighting and the calculation of the diffuse surface. The rest of the lighting elements were calculated using the Ray Tracing methods. The Radiosity method did not facilitate work on complicated and complex objects, so it was necessary to simplify the model. This led to the impoverishment of the stage and in the case of complex geometric objects, to distortions in the final image.

The aim of all methods was to reduce distortions in the rendered image and to get the shortest time of computation. The application of different models of light representation allowed the achievement of satisfactory results. The breakthrough was the introduction of the Photon Mapping method, which allows obtaining satisfactory results with reasonable computation time [3]. The algorithm gives the possibility to work in environments with more complex structures than it was using Radiosity, while being able to obtain the lighting effects, such as the Metropolis Light Tracing.

3 Photon Mapping

The original Photon Mapping algorithm, proposed by Jensen [4], consists of a two stage process in order to calculate Global Illumination. The first stage is based on emitting a defined number of photons from the light source and tracing them through the scene. The energy of single photon amounts (2)

$$E_p = \frac{1}{\mathrm{n}} E_s \tag{2}$$

where:

n - number of photons emitted from light source, E_s - integral energy of light source.

When the photon hits the scattering surface, information about the position of the hit point, photon energy and incoming vector, are saved into a spatial data structure called the photon map. If the surface is specular or refractive, the photon is propagated further in accordance with the law of reflection and Snell's law, until it hits the diffusive surface. Then, according to its coefficient of probability, the photon is either absorbed or reflected with a modified energy value. The process is repeated until it reaches a specified depth of recursion.

After the end of the propagation of photons, the second stage takes place. In order to render the scene using the information collected in the photon map, the classical Ray Tracing algorithm is performed. Rays are traced sequentially from the position of the virtual camera through each pixel of the image.

The places where rays intersect with the visible surfaces are marked as hit points. The radiance at the particular hit point can be computed using density estimation by querying for a defined number of nearest photons from the photon map. If the number of neighbouring photons is not sufficient, then the radius of the searching hemisphere is increased up to the pre-established maximum value. The final pixel color for the single hit point is calculated using the formula (3)

$$L(x,\omega) = \sum \rho_x(\omega,\omega') \frac{\Delta\Phi(x,\omega')}{\pi r_n^2} \tag{3}$$

where:

ρ_x- is the bidirectional reflectance distribution function (BRDF), Φ- is the energy of the incident photon, πr_n^2 - is the area of a circle with specified searching radius.

Undoubtedly, a great advantage of the photon mapping algorithm is a direct decoupling of the illumination calculations from the geometry of the scene, in contrast to the Radiosity. However, the practical application of the algorithm involves two basic limitations such as: limited size of the photon map and the time needed to search for a fixed number of neighboring photons. Some of the improvements aimed to eliminate these limitations are presented in the next section.

3.1 Optimizations of Photon Mapping

In order to improve the most time consuming phase of the algorithm, which is searching for adjacent photons, it is proposed that the photon map structure is reorganized into a K-d tree, thus greatly narrowing down the searching area.

Another improvement in order to speed up the calculations was suggested by Keller and Wald [5]. The previous procedure assumed using a standard Monte Carlo technique called Russian Roulette as a mechanism to determine whether the reflected photon should be absorbed or propagated further in the scene. This led to a completely random distribution of photons, without taking into account visibility from the camera viewpoint. Determining areas of the scene that are important in terms of image quality (for instance the accumulation of objects), allows using a projection map, warped onto a bounding shape for the light source, that causes emition of a greater number of photons towards specified directions of the scene. Projection maps also improved rendering of the caustics phenomena, which requires a substantial density of photons focused on refracting surfaces. However, this approach did not affect the reflected photons responsible for the indirect lighting. Therefore, the emission of photons from the source was preceded by sending rays from the viewpoint of the camera and defining visibility coefficient of the surface, which was taken into account when storing photons into the photon map. This allowed the achievement of much more accurate results with the same number of photons emitted.

One of the main limitations of the Photon Mapping algorithm is the assumption that the surface in the neighborhood of the estimated illumination point is planar. If the geometry of the environment is not flat or is significantly smaller than the radius of the searching hemisphere, then the equation (3) results in the appearance of artifacts in the form of underexposed areas. The algorithm proposed in [6], is an attempt to reduce these kind of errors, based on information about the geometry in the vicinity of the photon hit point. The next amendment was the implementation of an adaptive surface approximation method using a series of polygons, depending on the conditions (number of samples, as well as a discontinuity of surface illumination) (Fig. 1).

Fig. 1 Artifacts in the form of shadow leaks on the edges and corners (left) and their correction by adaptive surface approximation (right).

The results of the experiment [7] confirm that the adaptive method is more efficient than other previously used techniques in terms of both memory requirements and execution time. A heuristic approach to the same topic suggested by Tobler and Maierhofer [8] eliminates shadow leaks by replacing the estimations circle area with an octagon, or eight octagons in the case of corners of the walls. Thanks to the above solutions, it is possible to precisely calculate the indirect illumination near the edges and the relatively small isolated objects.

3.2 Reverse Photon Mapping

The concept of reverse photon mapping [9] consists of determining the points where light intensity will be calculated, before the propagation of photons in the scene. In the standard procedure this was done only during the second stage. Reordering the sequence of actions caused that instead of searching a given number of neighboring photons for each hit point, every photon colliding with a diffusive surface searches, in a restricted area, for hit points for which it contributes to the calculation of light intensity. This approach, defined as Ray Splatting, requires an appropriate data structure of hit points to perform an efficient searching method. Partially calculated radiance is stored in a radiance map consisting of several files with a final resolution of the rendered image.

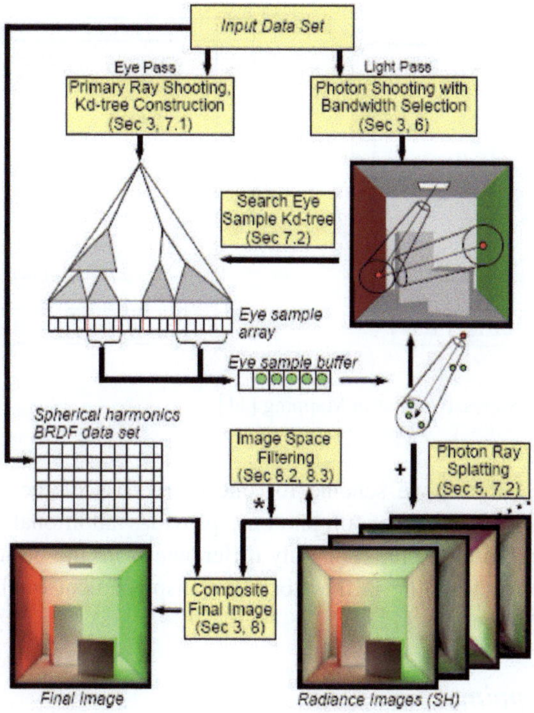

Fig. 2 Ray Splatting diagram [10]

3.3 Progressive Photon Mapping

In extreme cases, the limited capacity of photon maps prevents highly detailed rendering above a specified number of photons and a permitted number of reflections. Development of Reverse Photon Mapping allows for the progressive approximation of the light intensity, and thus eliminates the weakness of the standard Photon Mapping algorithm (which is the increasing demand for memory to store the information about gathered photons). Progressive photon mapping algorithm proposed by [11] consists of two passes. The Ray Tracing Pass detects intersections of the rays coming from the camera with the geometry of the scene and store information about position of the intersection point, assigned pixel, incoming vector, BRDF coefficient (Bidirectional Reflectance Distribution Function), actual searching radius and calculated radiance. The second stage called Photon Tracing Pass consists of a cyclic propagation of photons in the scene followed by the regular updating for each of the hit points approximation value of photons energy and decreasing searching radius.

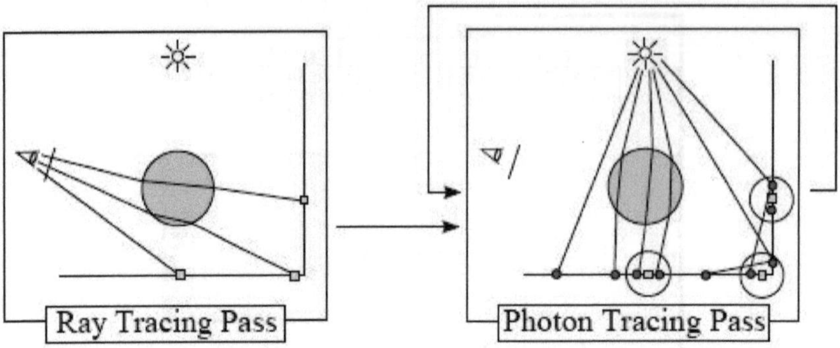

Fig. 3 Schema of Progressive Photon Mapping [11]

Supplementing the above scheme for one more pass of the Distributed Ray Tracing after each propagation of photons, provides additional effects such as depth of field and motion blur. The only difference from the basic version of the Progressive Photon Mapping is the stochastic displacement of the impact point within a fixed environment [12].

3.4 Rays Mapping

Another method based on Photon Mapping takes into account the paths of photons (identified with rays), rather than the impact points [13]. Estimation of the density of photons located on the flat surface at the sampling point is replaced by determination of rays intersecting particular searching area. An example of a ray map - a data structure storing the information on the transport of light in the scene was suggested by Havran [14]. With this approach, the assumption that the nearest environment of a hit point must be flat is not necessary for the proper calculation of light intensity. As a result, the algorithm is far more independent of the geometry of the scene than the standard Photon Mapping method. In order to determine the intensity, all rays passing through the searching disk with a set radius are included, not only the first n photons found. Comparing this two algorithms in terms of quality weighs strongly in favour of the Ray Mapping, however it needs far more computation time and memory resources.

A typical problem of this method is to identify rays intersecting with the concave parts of the grid (Fig. 4). The solution can be achieved by storing in the data structure information about both the first and second intersection with geometry. In this way it is possible to check whether the ray intersection with the surface of the searching disk occurred before second intersection with the scene geometry, if so then this is a concave area and for the correct calculation of light intensity considered ray also should be taken into account.

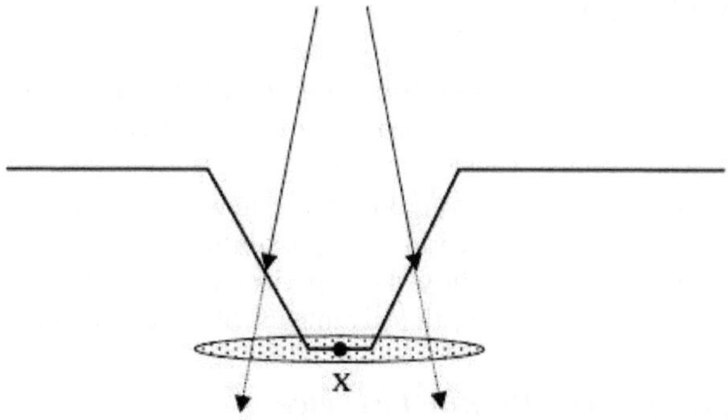

Fig. 4 Determination of radius for the concave parts of the grid [13]

The second common problem is the estimation of the density of rays passing through the surfaces of a disk around the sampling point located on the edge, or on crossing planes. In such cases, part of the disc will not contain any intersections (Fig. 5), so the sum of intensity will be divided by a too large area value, resulting in darker areas. The solution is to cut the surface of disks by using the planes designated by the vectors perpendicular to the normal at the edges. Naturally, this cause an additional burden in time.

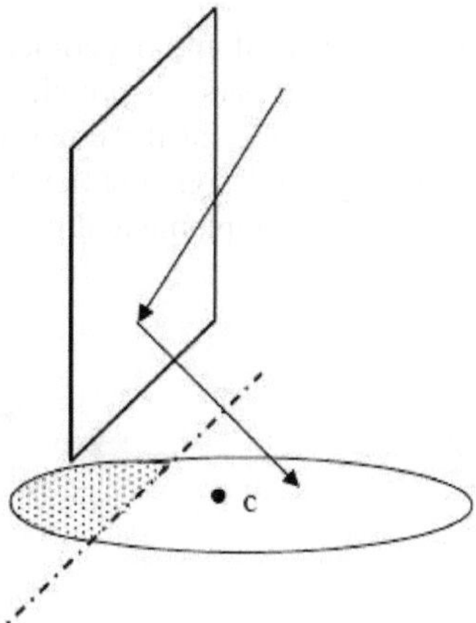

Fig. 5 Area of the disk without intersecting rays [13]

The Ray mapping algorithm requires multiple, time-consuming calculation of the intersection ray with the disc. Optimizations in the way of identifying the most probable set of rays (with convergent directions) and spatial indexing disks that are surrounded by the sampling point have been proposed by [15].

4 Rendering System

There are lots of commercial applications for the photorealistic rendering using global illumination methods. The most popular rendering engines are Mental Ray, V-Ray and Maxwell Render. These solutions are based on a standard photon mapping method. Research on the development the new algorithms is conducted in many scientific centers over the world. The possibility to share between other institutions and researchers would allow a comparison of algorithms used to develop new solutions and methods that use the advantages of different algorithms.

4.1 Description of the Planned Service

The goal of our project is to create a rendering system based on the latest developments in the field of photorealistic computer graphics. Implementation of algorithms based on the Photon Mapping will help to create effective tools for data visualization. Appropriate organization of the code; give the possibility to share information between researchers. The system will design, that each user can develop an explicit algorithm or implement their own solutions. Use of the net-work should be not only in terms of the exchange of data but also detailed documentation describing the used solutions. The program will be used not only by developers but also by users who wish to perform visualization of 3D models.

4.2 Structure of the Renderer

The application should allow the use of mesh geometry, defined by open data formats (3ds, obj). For this purpose, a parsing module for popular geometry formats should be implemented. The project of the component of the rendering engine will consist of a part enabling the design and development of algorithms and components enabling the testing of the implemented methods.

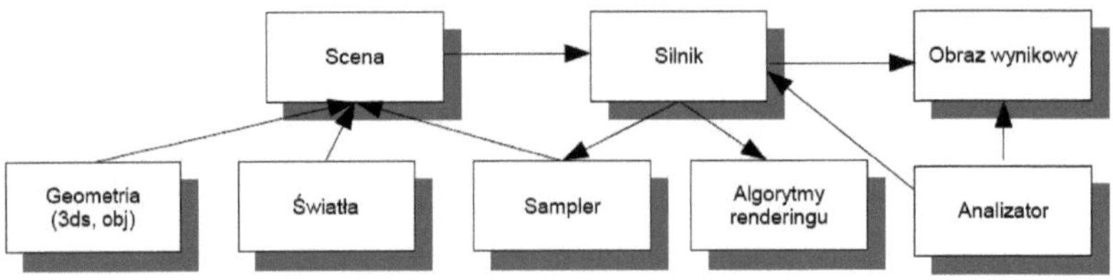

Fig. 6 Diagram of rendering system

After successfully launching the service with the basic photon mapping algorithm, it is planned to further implement the following modifications to the method described in section 3. This will provide a new method or enrich existing methods based on existing development trends.

5 Summary

Global illumination methods are the current topic for many research units in the country and abroad. Despite the existence of a number of tools that use the solutions developed, there lacks a system of allowing the sharing of drop algorithms. It therefore seems reasonable to create a system of an educational and research character, that would be helpful the development of global illumination algorithms in the future.

References

[1] Kajiya James, T.: The rendering equation. In: Proceedings of SIGGRAPH 1986, vol. 20(4), pp. 143–150 (1986)

[2] Dutré, P., Bekaert, P., Bala, K.: Advanced Global Illumination. AK Peters (August 2003), ISBN 1-56881-177-2

[3] Jensen, H.W.: Global illumination using photon maps. In: Rendering Techniques 1996, pp. 21–30. Springer (1996)

[4] Jensen, H.W.: Realistic Image Synthesis using Phothon Mapping. AK Peters (2001)

[5] Keller, A., Wald, I.: Efficient Importance Sampling Techniques for the Photon Map. In: Proceedings of Vision Modelling and Visualisation 2000, Saarbruecken, Germany, pp. 271–279 (2000)

[6] Hey, H., Purgathofer, W.: Advanced radiance estimation for photon map global illumination. Computer Graphics Forum 21(3), 541–546 (2002)

[7] Wong, K.W., Wang, W.: Adaptive density estimation using an orthogonal series for global illumination. Computers & Graphics 29(5), 738–755 (2005)

[8] Tobler, R.F., Maierhofer, S.: Improved illumination estimation for photon maps in architectural scenes. In: Proc. of WSCG, pp. 257–262 (2006)

[9] Havran, V., Heezog, R., Seidel, H.-P.: Fast final gathering via reverse photon mapping. Proceedings of Eurographics on Computer Graphics Forum 24, 3 (2005)

[10] Herzog, R., Havran, V., Kinuwaki, S., Myszkowski, K., Seidel, H.-P.: Global illumination using photon ray splatting. Comput. Graph. Forum 26(3), 503–513 (2007)

[11] Hachisuka, T., Ogaki, S., Jensen, H.W.: Progressive photon mapping. ACM Transactions on Graphics (SIG- GRAPH Asia Proceedings) 27, 5, Article 130 (2008)

[12] Hachisuka, T., Jensen, H.W.: Stochastic Progressive Photon Mapping. ACM Transactions on Graphics (SIGGRAPH Asia 2009) (2009)

[13] Lastra, M., Urena, C., Revelles, J., Monyes, R.: A particle-path based method for Monte Carlo density estimation. In: Proc. of the 13th Eurographics Workshop on Rendering, pp. 33–40 (2002)

[14] Havran, V., Bittner, J., Herzog, R., Seidel, H.P.: Ray maps for global illumination. In: Proc. of Eurographics Symposium on Rendering, pp. 43–54 (2005)

[15] García, R.J., Ureña, C., Revelles, J., Lastra, M., Montes, R.: Density estimation optimizations for global illumination. In: WSCG 2006 Short Communications Proceedings (2006)

[16] Stolarek, J., Lipinski, P.: Improving Digital Watermarking Fidelity Using Fast Neural Network for Adaptive Wavelet Synthesis. Journal of Applied Computer Science 18, 61–74 (2010)

A Compiler for a Domain-Specific Language for Rapid Implementation of DSP Transforms and Filter Banks

Marek Parfieniuk, Marcin Kazberuk, and Kamil Kowalewski

Abstract. Transforms and filter banks are widely used in digital signal processing (DSP), but not many specialized tools exist for developing such systems. Moreover, these tools are somewhat user unfriendly and lack useful features, and their drawbacks cannot be easily solved. Therefore, we have recently proposed a domain-specific language, called TDL, for describing systems clearly, compactly, and quickly. In this chapter we present an associated compiler, which converts TDL descriptions into Java/C++ implementation code, so as to save programming effort. In particular, the dataflow and data structures are discussed that are related to code generation.

1 Introduction

Filter banks and transforms are widely used in digital signal processing (DSP), as a means of subband decomposition of signals [1–3]. The most prominent representatives of them are the Fast Fourier Transform (FFT), the Discrete Cosine Transform (DCT), and the Discrete Wavelet Transform (DWT). The first is the fundamental tool of signal analysis, the second is used in the JPEG and MPEG-2 standards for image and video coding, respectively, whereas the last one forms the basis of the multiresolution approach to signal processing, which is employed e.g. in the JPEG2000 standard for image compression.

Three research directions can be identified in the related field of science. Firstly, novel filter banks are designed that are optimized to suit a particular application or signal class. Secondly, improved implementations of known systems are developed that are more computationally efficient or more robust to finite-precision effects in the case of fixed-point implementation [4]. And finally, new applications of filter banks and transforms are regularly proposed.

Marek Parfieniuk · Marcin Kazberuk · Kamil Kowalewski
Faculty of Computer Science, Bialystok University of Technology
Wiejska 45a, 15-307 Bialystok, Poland
e-mail: m.parfieniuk@pb.edu.pl

P. Lipiński, K. Świrski (Eds.): Towards Modern Collaborative Knowl. Sharing Syst., SCI 401, pp. 83–100.
springerlink.com

Obviously, for persons faced with any of the aforementioned tasks, it is advantageous to be able to implement a transform quickly and to rapidly prototype a program that uses or evaluates it. However, developing filter banks using the existing tools is generally tedious and time consuming.

In order to solve this issue, one of the authors has recently introduced a domain-specific language (DSL) called the Transform Description Language (TDL), which allows transforms to be described clearly, compactly, and quickly [5]. The advantages are related to its syntax, which tightly links TDL code to the signal flow graph of a system. The author has also developed a prototype compiler which converts simpler TDL descriptions into Java implementation code.

Improved and extended versions of both language and compiler are the topic of our project of creating an application for the TEWI platform. The goal is to develop a functional tool, which allows persons interested in implementing transforms or filter banks to save time and effort.

This chapter describes our TDL compiler from both functional and programming points of view. We present our design assumptions as well as the compilation flow and data structures that we have developed. In order to make the paper self-contained, the TDL language is also described, and similar existing tools are briefly reviewed.

2 TDL: A Domain-Specific Language for Describing Transforms and Filter Banks

2.1 Motivation for Developing the Language

Transforms and filter banks decompose signals into frequency subbands [1]–[3]. Signal samples are converted into coefficients that are less correlated and thus better suited to compression, noise reduction, pattern recognition etc.

From this point of view, such systems are usually identified with sets of independent filters, as illustrated in Figure 1a. In the case of designing or implementing a computational scheme, however, it is more convenient to use the equivalent scheme in Figure 1b, where the input vector is multiplied by the so-called polyphase matrix which describes the system. Block transforms are determined by scalar matrices, whereas polynomial matrices describe filter banks, also called lapped transforms.

System optimization and adaptive processing require filter coefficients to vary during program execution. This is achieved by making matrix elements dependent on a smaller number of system coefficients, which are controlled by an external algorithm. Such dependence is realized by representing the polyphase matrix as a product of simpler factors, some of which are directly determined by the coefficients. These simpler matrices have specific forms and can be identified with elementary transformations [1,4,6,7]: rotations, reflections, lifting steps, quaternion multiplications, gains, and delays.

Computations can also be saved by ingeniously factorizing the polyphase matrix. Therefore transforms are commonly implemented as arrangements of the aforementioned building blocks, without explicitly computing matrix elements.

As readers unfamiliar with transforms and matrix factorizations might expect an example, the Loeffler scheme for computing the 8-point DCT [8] is shown in Figure 2a. The algorithm requires 26 additions and 16 multiplications, which is several times less operations than when computing the output vector X directly

$$X_m = \frac{1}{2}\sum_{n=0}^{7} \underbrace{s_m \cos\left(\frac{\pi}{16}(2n+1)m\right)}_{m,n\text{-th element of the DCT matrix}} x_n, \qquad m = 0\ldots7, \quad s_0 = \frac{1}{\sqrt{2}}, \quad s_{1,\ldots,7} = 1, \tag{1}$$

i.e. by multiplying the input vector x by the 8×8 DCT matrix.

From our point of view, it is crucial that factorizations define signal flow graphs in which data move from input to output without feedback, through a series of stages, each of which comprises a single elementary transformation, or more generally a subtransform. If some subtransforms are computed in parallel, they can be arbitrarily assigned to stages, but the approach is still valid.

It can always be assumed that a subtransform has the same number of inputs and outputs, and that it is connected to some outputs of the subsystem composed of all preceding stages, not affecting the number of the outputs.

Both number and arrangement of building blocks are fixed. There are neither optional subtransforms nor switched connections. Only coefficients of elementary transforms can be modified from outside the transform, so as to change the way of processing input data.

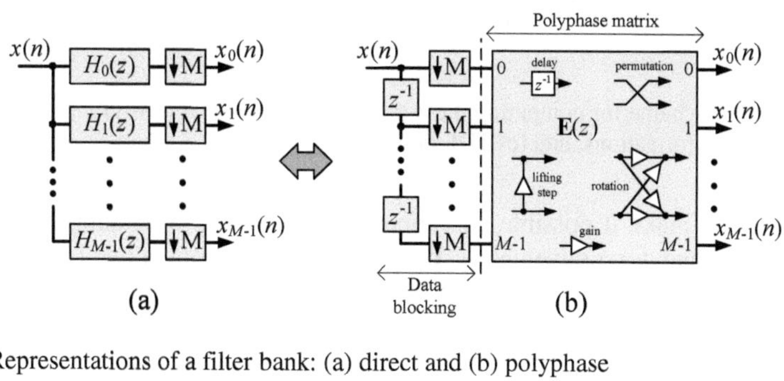

Fig. 1 Representations of a filter bank: (a) direct and (b) polyphase

(a)

(b)

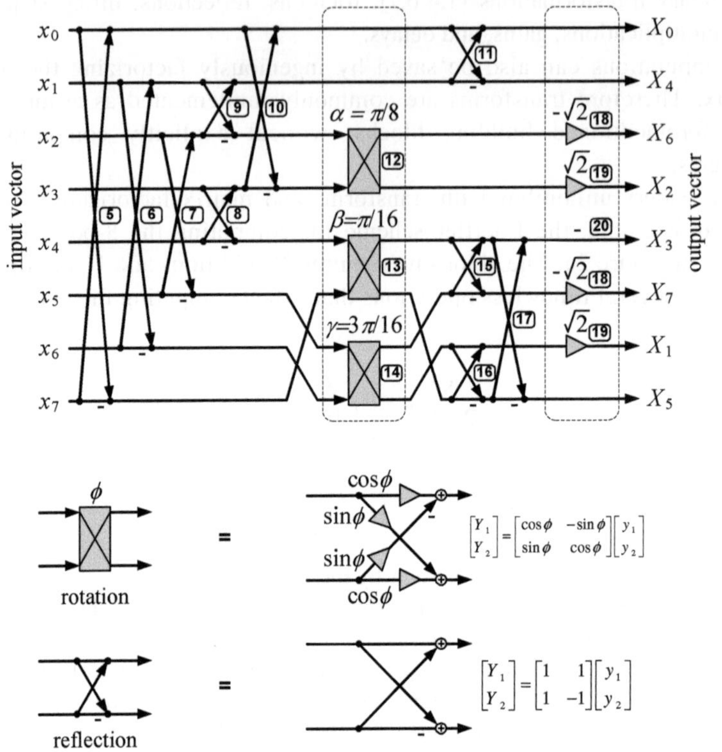

(c)

```
transform Loeffler<8,0> use
    reflection = [1,  1;
                  1, -1];
do
    [0, 7] @ reflection;
    [1, 6] @ reflection;
    [2, 5] @ reflection;
    [3, 4] @ reflection;
    [1, 2] @ reflection;
    [0, 3] @ reflection;
    [0, 1] @ reflection;
    [2, 3] @ rotation(pi/8);
    [4, 7] @ rotation(pi/16);
    [5, 6] @ rotation(3 * pi/16);
    [4, 5] @ reflection;
    [6, 7] @ reflection;
    [4, 7] @ reflection;
    [2, 5] @ -sqrt(2);
    [3, 6] @ sqrt(2);
    4 @ -1;
    0 : 7 @ permute([0,4,6,2,3,7,1,5]);
end
```

Fig. 2 Loeffler scheme for computing the 8-point DCT: (a) signal flow graph, (b) details of elementary transformations, and (c) TDL description

These facts make it possible and appealing to develop a domain-specific declarative language for conveniently describing such computational schemes. Its syntax should allow developers to declare the number of system inputs and to specify consecutive stages of a scheme, using both built-in elementary transformations and customized subtransforms. A compiler for such a language, able to convert descriptions into C/C++/Java code, would be a nice tool for implementing transforms and filter banks.

(a)

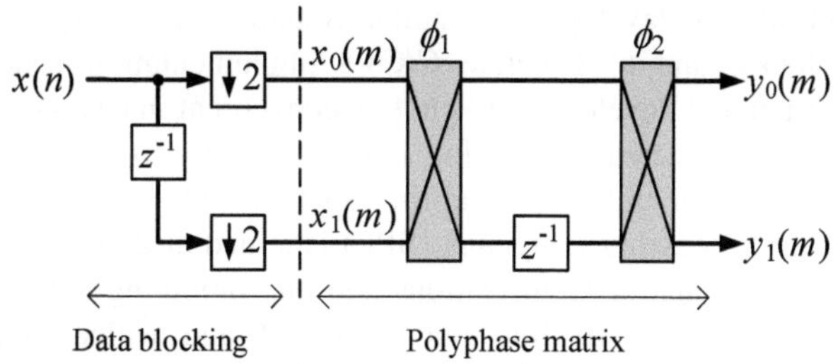

(b)

```
transform LatticeFilterBank<2, 2> use
    rotation(phi) = [cos(phi), -sin(phi); sin(phi), cos(phi)];
do
    : @ rotation($0);
    2 @ delay(1);
    : @ rotation($1);
end
```

Fig. 3 Two-channel paraunitary filter bank: (a) lattice structure [1] and (b) TDL description

With this idea in mind, one of the authors has introduced the TDL language [5], whose syntax allows transforms and filter banks to be clearly and compactly described in terms of elementary transformations. This is achieved by requiring developers to specify for each stage both the indexes of the outputs of preceding stages that have to be processed and the subtransform to be computed on these outputs.

Examples of TDL descriptions are shown in Figures. 2 and 3. In Figure 2, the numbers in the boxes indicate lines of the description that correspond to the subtransforms.

2.2 Comparison to Existing Tools

In spite of the significant role that transforms play in DSP, not many specialized tools exist for developing such algorithms. To the best of authors' knowledge, the only exceptions are MATLAB, Simulink and SPL/SPIRAL [9, 10], which however have some shortcomings.

Supporting matrix computations, MATLAB seems naturally suited to studying transforms. However, the tool is not as useful as it could be, even though it is widely used to develop DSP algorithms. Especially, modeling matrix factorizations requires both unclear and inefficient MATLAB code, behind which are excessive data movements and unnecessary computations, which are related to multiplying sparse matrices and can be only partially avoided by complicated array indexing. Modeling polynomial matrices is even more troublesome as it requires workarounds like using loops or objects. Finally, it seems generally impossible to easily convert a MATLAB script into efficient code in another programming language.

Simulink allows developers to model systems graphically, using a rich set of building blocks, and to generate C/C++ implementation code from schemes. However, graphical development is not so convenient in cases of many blocks and interconnections, whereas neglecting Java is contrary to recent trends in the software engineering for embedded real-time systems [11, 12].

The Signal Processing Language (SPL) is aimed at describing DSP algorithms that are based on matrix factorizations and on computing matrix-by-vector products. Even though it is flexible and functional, and a nice compiler is publicly available, a closer look reveals two issues that can be annoying for developers. Firstly, the SPL syntax is based on Polish notation and thus is unclear and unnatural for persons using modern programming languages, and secondly, the compiler is only able to generate FORTRAN or C code.

The SPL is a part of the SPIRAL project, which is aimed at developing tools that automatically generate machine-level implementations of linear, transform-based DSP algorithms, optimizing implementations so as to speed up computations by exploiting features of modern microprocessors [10]. As only tools and results related to blocks transforms and simple filters are publicly available, it seems impossible to evaluate how well SPIRAL helps in developing filter banks.

In our opinion, the aforementioned tools do not allow for describing transforms as clearly, efficiently, and compactly as possible, and this problem cannot be solved without radically changing syntax. Additionally, Java code could be generated. Finally, the MATLAB and Simulink are expensive to buy.

The above facts justify our idea of developing a new language and compiler that make implementing transforms easier and less time consuming. However, the reader might wonder if creating a specialized Java or C/C++ library would be similarly helpful, requiring less programming effort. According to our experience, developing and using a library is not much easier and has evident shortcomings. Firstly, library function calls cannot be as natural, clear, and compact as expressions in a dedicated language. Secondly, the most effective code optimization strategies cannot be employed: computing partial results in advance, avoiding both procedure calls and memory allocation, and loop unrolling. And finally, it is impossible or computationally costly to prevent bugs related to passing semantically incorrect arguments to a function.

2.3 Structure of Description Files

The properties of transforms and filter banks, which we have described in Section 2.1, can be directly exploited to describe systems clearly, compactly, and easily. In particular, they allow the TDL language to be declarative. Thus, it lacks procedural programming features: conditional statements, comparison operators, loops, types, variables, and even functions in the usual sense.

The general structure of TDL descriptions can be presented using the examples in Figures 2 and 3. A description begins by specifying an identifier of the transform, number of inputs, and number of system coefficients. Then, the optional section follows in which matrices and subtransforms can be defined in order to save effort, prevent bugs and clarify code. Finally, consecutive stages of the computational scheme are described, beginning with the stage closest to the system inputs. Initially, the system comprises the single stage that simply moves data from inputs to outputs.

A stage description begins by specifying the indexes of some outputs of the subsystem composed of all preceding stages. Then, separated with the "@" character, a subtransform is specified that has to be applied to these outputs, so as to form an extended subsystem. Such an approach allows elementary transforms to be incorporated into computational schemes without the necessity of constructing sparse matrices or complicated array indexing, which are unavoidable in both SPL and Matlab, respectively.

A square matrix can often serve as a subtransform specification. In this case, a linear transformation of input data is declared that is determined by the matrix. Elements of the matrix can be directly specified using matrix notation, possibly referring to a build-in matrix or a matrix of the definition section.

Stages can also refer to subtransforms defined by a developer. Subtransforms are specified using the same means as the main transform. In particular, they can contain low-level subtransform and matrices.

Output indexes and operation arguments can be specified by directly writing matrix expressions or by referring to matrices of the definition section. Additionally, we defined two simplifications: skipping output specification is equivalent to using that of the preceding stage, whereas writing only a colon is equivalent to selecting all outputs.

2.4 Syntax of Matrix Expressions

In the TDL, matrix expressions can be written using a subset of the MATLAB syntax. In particular, the most practical operators have been adopted that are responsible for the compactness and naturalness of MATLAB scripts, i.e. apostrophe denotes matrix transposition, brackets – concatenation, and colon – number sequence. This should allow persons familiar with MATLAB to learn and use TDL without much effort.

We have omitted MATLAB features that are useless in describing transforms, namely functions and language constructs related to matrix analysis, extracting submatrices, and procedural programming. Instead, built-in functions have been defined for generating matrices of elementary transformations that are most often used in the computational schemes of interest.

Table 1 shows examples of matrix expressions that can be used in TDL descriptions. A notable difference from MATLAB is that filters can be matrix elements. Thus, the proposed language unifies modeling of both filter banks and block transforms, which are described by polynomial and scalar matrices, respectively. It is also noteworthy that system coefficients are referenced using the dollar symbol "$".

Table 1 Examples of TDL matrix notation

Notation	Description
`liftu(a)`	lifting step: $\begin{bmatrix} 1 & a \\ 0 & 1 \end{bmatrix}$
`[matrix ; ... ]`	vertical concatenation of matrices
`fliplr(matrix)`	horizontal (left-for-right) flip
`zeros(m, n)`	$m \times n$ matrix of zeros
`matrix .* matrix`	multiplying matrices element-by-element
`start : step : stop`	number sequence
`matrix'`	transpose
`permute(3,1,4,2)`	permutation matrix: $\begin{bmatrix} 0 & 0 & 1 & 0 \\ 1 & 0 & 0 & 0 \\ 0 & 0 & 0 & 1 \\ 0 & 1 & 0 & 0 \end{bmatrix}$
`J(3)`	3×3 counter-identity matrix: $\begin{bmatrix} 0 & 0 & 1 \\ 0 & 1 & 0 \\ 1 & 0 & 0 \end{bmatrix}$
`refl45n()`	normalized reflection by 45°: $\begin{bmatrix} 1 & 1 \\ 1 & -1 \end{bmatrix}$
`refl(c, s)`	reflection: $\begin{bmatrix} c & s \\ s & -c \end{bmatrix}$
`rota(a)`	planar rotation by a given angle: $\begin{bmatrix} \cos(a) & -\sin(a) \\ \sin(a) & \cos(a) \end{bmatrix}$
`sin(matrix)`	sine of matrix elements
`$n`	n-th system coefficient
`delay(n)`	n-sample delay
`fir(coefficients)`	FIR filter
`quatl(a, b, c, d)`	left quaternion multiplication matrix: $\begin{bmatrix} a & -b & -c & -d \\ b & a & -d & c \\ c & d & a & -b \\ d & -c & b & a \end{bmatrix}$

3 The TDL Compiler

3.1 Design Requirements and Assumptions

A TDL compiler should allow applications to be developed using the workflow in Figure 4. The tool converts an implementation-independent TDL description of a transform into Java or C/C++ source code according to preferences specified in a separate file. In order to assembly an application, a developer attaches the resulting files to the rest of its source code and uses a platform-specific compiler to produce executables. Separating code generation options from descriptions allows files of both kinds to be reused in various projects.

By using TDL, developers can avoid both tedious programming and some bugs. Nevertheless, they are responsible for describing systems correctly and lucidly, i.e. using optimal TDL constructs. This is necessary because the TDL compiler optimizes generated code using only simple yet effective strategies: data movements and subroutine calls are avoided, loops are unrolled, and partial results are computed in advance. Assuming that the Java/C/C++ compiler is responsible for low-level code optimization, quite efficient implementations of transforms can be obtained.

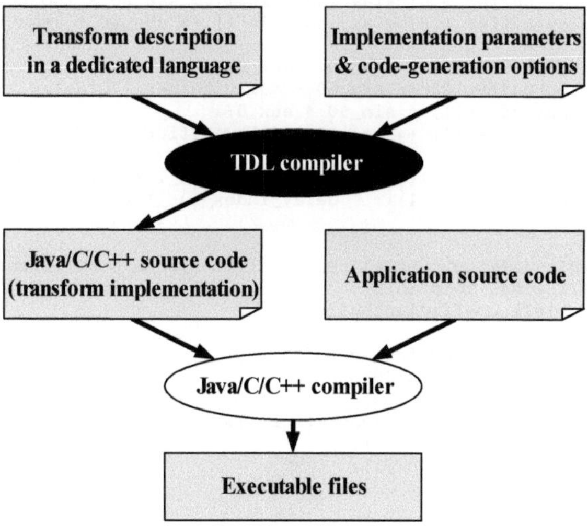

Fig. 4 TDL-based development of applications

On the other hand, implementation code is made accessible by generating descriptive identifiers and comments, and by not using nonstandard libraries. This should allow generated source files to be customized by hand and used in ways that cannot be predicted at the time of developing a compiler. In most cases, developers should be able to adjust implementation details by only setting code generation preferences: data types (fixed- vs. floating-point numbers) and access methods (pointers vs. one- and two-dimensional arrays) to the input, output, and coefficient buffers.

```
package transform;

import java.util.Arrays;

public final class MyTransform {
  private int delay_index_in_1, delay_index_out_1;
  private final double[] delay_buffer_1_0 = new double[2];

  private double cos_$0;
  private double sin_$0;
  private double cos_$1;
  private double sin_$1;

  public final void process(
    double[] inputBuffer,
    double[] outputBuffer
  ) {
    double aux_0;

    aux_0 = cos_$0 * inputBuffer[0];
    aux_0 -= sin_$0 * inputBuffer[1];
    delay_buffer_1_0[delay_index_in_1] = sin_$0 * inputBuffer[0];
    delay_buffer_1_0[delay_index_in_1] += cos_$0 * inputBuffer[1];

    outputBuffer[0] = cos_$0 * aux_0;
    outputBuffer[0] -= sin_$0 * delay_buffer_1_0[delay_index_out_1];
    outputBuffer[1] = sin_$0 * aux_0;
    outputBuffer[1] += cos_$0 * delay_buffer_1_0[delay_index_out_1];

    delay_index_in_1++;    delay_index_in_1 %= 2;
    delay_index_out_1++;   delay_index_out_1 %= 2;
  }

  // Initializing partial results from system coefficients
  public final void init(
    double[] coefficientBuffer
  ) {
    double $0 = coefficientBuffer[0];
    cos_$0 = Math.cos($0);
    sin_$0 = Math.sin($0);
    double $1 = coefficientBuffer[1];
    cos_$1 = Math.cos($1);
    sin_$1 = Math.sin($1);
}

  // Initializing delay buffers
  public final void resetDelays() {
    delay_index_in_1 = 1;
    delay_index_out_1 = 0;
    Arrays.fill(delay_buffer_1_0, 0);
  }
}
```

Fig. 5 Implementation code for the filter bank of Figure 3

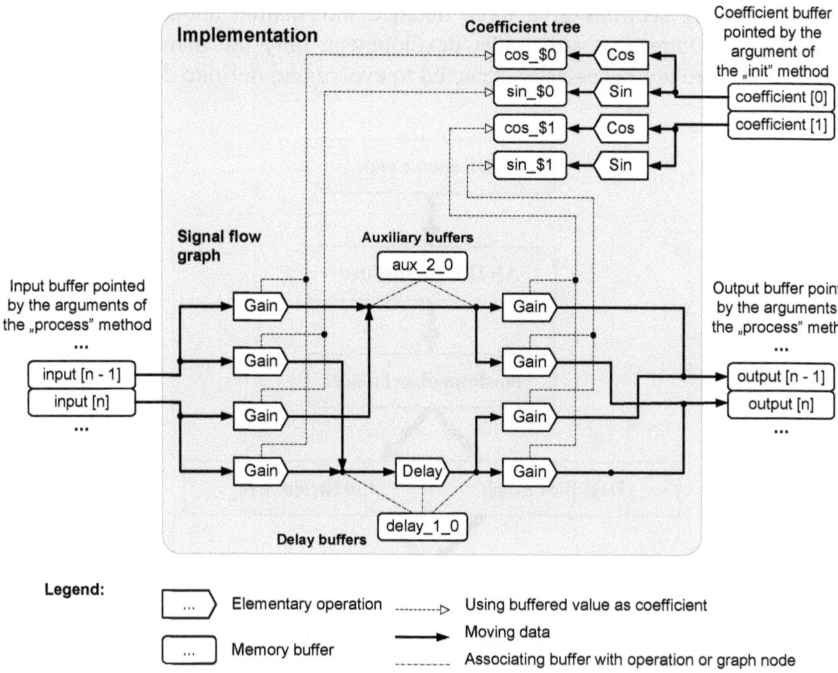

Fig. 6 Implementation architecture for the filter bank of Figure 3

A transform implementation comprises three functions, which cooperate using shared variables or being methods of the same class. The "resetDelays" function clears data buffers used to implement delays and filters. The "init" function precomputes coefficients of elementary operations, based on passed values of system coefficients. Finally, the "process" function is responsible for computing the output vector for a given input vector, using precomputed coefficients as well as current contents of data buffers.

The "process" function has to be repeatedly called by higher-level application code, each time with different input data. Selecting successive input vectors from a long sequence of input samples can easily be realized without data copying, by additionally passing to the function an offset into the input buffer.

The proposed architecture of transform implementation can be evaluated using Figures 5 and 6.

3.2 General Flow of Data within the Compiler

The TDL compiler we are developing as an application for the TEWI platform is written in Java. Its source code consists of more than 100 classes, which implement the general dataflow of Figure 7. Three intermediate representations of a transform are created one by one in order to convert a TDL description into implementation source code. This stratification is necessary for satisfying some functional requirements as well as for simplifying development and testing.

The following sections give more detailed information about all representations. As the compiler is still under development, only the main ideas are presented, which are not necessarily expected to evolve into definite design decisions.

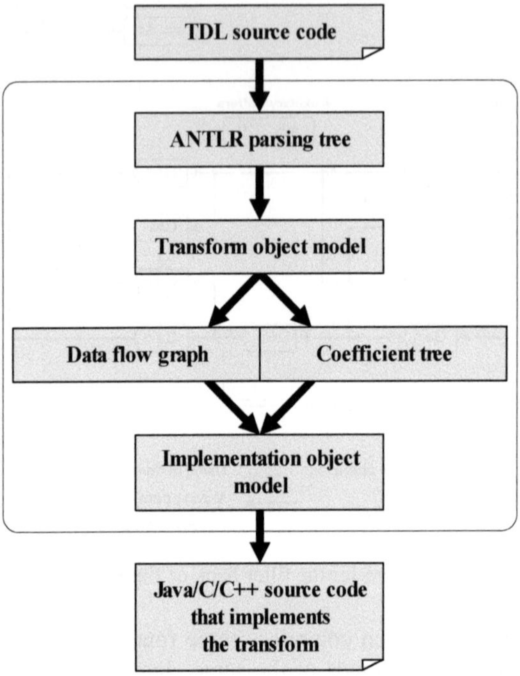

Fig. 7 Flow of data within the TDL compiler

3.3 ANTLR-Based Parser

The essential parts of each compiler are lexical analyzer and parser. The former converts a stream of source code characters into a higher-level stream of tokens: identifiers, keywords, numeric literals etc. Then the latter matches token sequences to expressions in a given language. If a valid sentence is detected, then appropriate actions related to language translation are executed. If tokens appear in an inappropriate order, then a syntax error is signaled.

In the previous version of our TDL compiler, the lexer-parser module was written by hand. In the present version, it is developed using ANTLR, a state-of-art parser generator, which is widely used by DSL developers [13].

Given a language grammar, ANTLR produces a class library for translating code, which can be easily incorporated into a software project. In addition to allowing actions to be assigned to grammar elements, the tool can automatically create an Abstract Syntax Tree (AST), which is a hierarchical representation of source code, and tree parsers, which facilitate multi-pass translation. It also provides an interface to StringTemplate, a library for generating structured text, which can be used to simplify the final stage of code generation. All these mechanisms are based on assigning processing rules to grammar elements.

We have concluded that ANTLR facilities related to advanced language translation and code generation are not very useful in our project. The main problem is that higher-level processing of syntax trees requires complicated rules, which make grammar files intricate and difficult to develop. Moreover, they slow down ANTLRWorks, a GUI-based environment which makes editing and debugging grammars easier by highlighting syntax, checking errors, visualizing diagrams, etc.

AST construction operators and rewrite rules are the only ANTLR mechanisms we employ. They allow the tree to be optimized by omitting unnecessary symbols, like parentheses and semicolons, and to be made uniform by adding imaginary nodes and replacing tokens in cases when the same meaning can be expressed using different TDL constructs. For example, the language allows matrix addition to be specified either by using the "+" operator or by calling the "add" function, but there is no difference between these two from the compiler point of view. Even limited usage of relatively simple rules obscures grammar definitions, as illustrated in Figure 8, and considerably slows down ANTLRWorks.

```
additivex
    :
    (multiplicativex -> multiplicativex)
    (
        '+' t = multiplicativex -> ^(REFERENCE ID["add"] $additivex $t)
    |
        '-' t = multiplicativex -> ^(REFERENCE ID["sub"] $additivex $t)
    )*;
```

Fig. 8 Converting addition and subtraction into function calls by extending an ANTLR grammar with rewrite rules

3.4 Object Model for Transform Descriptions

The AST of our compiler is based on the default node implementation provided by ANTLR, namely „CommonTree". Using such an AST as a model for systems is not very advantageous, as parent-child relations among nodes as well as the related methods of the "Tree" interface are very general. It is desirable to represent a transform more specifically and naturally, namely as a sequence of stages, which can refer to matrices and transforms of the definition part of a description. This is why we have developed the object model shown in Figure 9. It is produced from the AST, after creating a complete tree by the parser, by traversing it in a depth-first fashion.

Half of the classes are related to identifying named (and possibly parameterized) transforms and matrices with both symbols and scopes in which symbols (and parameters) can be defined and used. This is necessary because the TDL language allows subtransforms and auxiliary matrices to be defined inside a transform.

The model for transform descriptions is based on even more complicated object model for matrix expressions, which are used to specify both arrangement and coefficients of subtransforms. Figure 10 presents the related classes, whose instances can be arranged into trees in accordance with TDL constructs to be compiled. Inheritance is widely used to simplify programming.

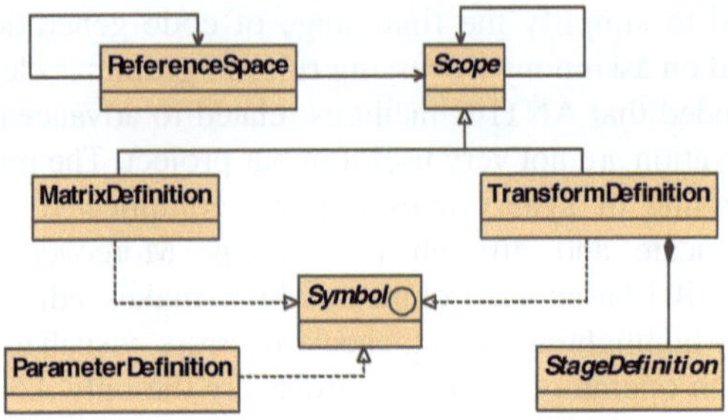

Fig. 9 Simplified UML class diagram for transform descriptions

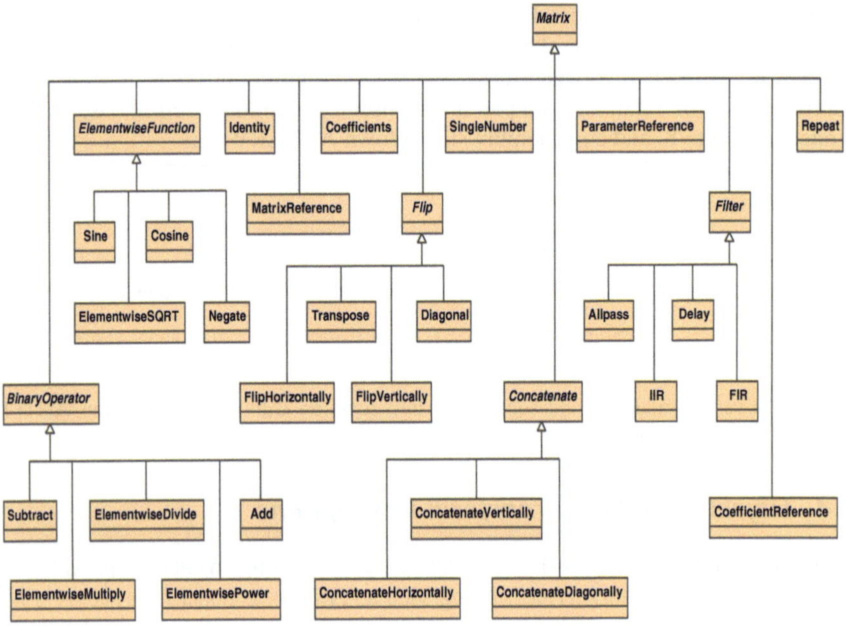

Fig. 10 Simplified UML class diagram for matrix expressions

3.5 Dataflow Graph and Coefficient Tree

The object model for transform descriptions represents dataflows only roughly, in terms of matrix processing. Our approach to code generation requires fine-grained modeling of both operations performed by a subtransform on its input data and transformations of system coefficients into coefficients of these operations. Therefore, the object model of a transform is further converted into two tightly coupled data structures.

The first one is the dataflow graph, whose edges represent elementary operations: gains, delays, filters etc., which can be directly expressed in a procedural language. Source and destination data buffers are assigned to every edge. Interconnected edges refer to the same buffer.

Results of some elementary operations depend on one or more coefficients. The latter can be not only constants but also functions of system coefficients. Thus, an additional data structure, a coefficient tree, is necessary, which describes how to derive operation coefficients from system coefficients. The roots of the tree are system coefficients, whereas the remaining nodes represent results of essential computations (arithmetic operations and basic functions) that can be performed on parent nodes and constants.

Obviously, edges of the dataflow graph refer to nodes of the coefficient tree, as illustrated in Figure 6. Both data structures can be quite easily generated from the object models of a given TDL description, mainly by using the correspondence between matrices and graphs, which is explained in Figure 2b.

Before creating a new node in the coefficient tree, it is checked if a node exists that represents a given expression for coefficient value. Thus, graph edges share tree nodes if this is possible.

3.6 Code Generation

Code generation begins by analyzing implementation preferences, determining an appropriate structure of implementation code, and preparing function headers. Then, the main two phases follows.

Firstly, the coefficient tree is traversed in order to generate the "init" function. Going from its roots to leaves, a tree walker produces source code that computes more and more complicated functions of system coefficients. Their results are assigned to persistent or auxiliary variables, depending on whether or not the corresponding tree nodes are referenced from within the dataflow graph. If the value related to a node serves as coefficient of one or more elementary operations, it is assigned by the "init" function to a static variable or class attribute, which allows it to be later, inside the "process" function, used in data processing. In order to make the latter possible, the node is updated with the identifier of the variable or attribute. If a node represents only an intermediate step in coefficient derivation, the related result is only used locally in the "init function".

Then, the consecutive edges of the dataflow graph are processed starting from the system inputs in order to generate the "process" function. For each edge, C/Java instructions are generated that read data from its source buffer, appropriately process them using coefficients prepared by "init", and store or add the result to the destination buffer, depending on whether or not the buffer has already been initialized by another operation. Obviously, for dataflows related to filtering, data structures necessary to buffering samples are defined before use.

Source code of different functions and their parts: headers, declaration sections, and instruction blocks, is accumulated in different text buffers, as separate code chunks. This is necessary because language constructs must be generated in order different from that in which they appear in a source file. Thus, the complier finishes its job by combining all chunks of implementation code into valid files.

Code generation is based on our own classes for storing text, generating identifiers, and managing code chunks, which can be identified with an implementation object model in Figure 7. It is also noteworthy that multiplication/addition operands are analyzed so as to omit unnecessary operations or replace them with simpler ones if possible. On the other hand, in our approach, no loops are generated even though arrays are processed. This can be viewed as loop unrolling, which is recommended by processor vendors as an optimization technique.

3.7 Integration of the Compiler with the TEWI Platform

The compiler will be made accessible to users of the TEWI platform in two ways. Firstly, they will be able to download an executable file to their workstations in order to locally process TDL descriptions. Secondly, using the JavaServer Pages (JSP) technology, a dynamic Web page has been developed that allows for uploading files to a server, compiling them remotely, and then downloading compilation results. The latter approach facilitates group work and ensures that the most recent version of the compiler is used. The layout of the Web page is shown in Figure 11.

(a)

(b)

Fig. 11 Web page for compiling TDL descriptions remotely: (a) list of compilation tasks of a user and (b) task details

The entry point to both executable download and remote compilation will be a project Web page that presents the TDL language and provides user guides.

4 Summary

Defining a grammar of a domain-specific language is much easier than engineering a compiler. In order to obtain a procedural implementation of a transform or filter bank from a high-level TDL description, several intermediate representations of the system must be created, each based on advanced algorithms and data structures.

In our project, ANTLR is only useful for developing a compiler front-end. Employing its facilities for code generation seems not very advantageous when input and output languages are diametrically different, so that complicated processing rules are necessary.

The language and compiler we have developed are an interesting alternative or supplement to the existing tools for implementing transforms and filter banks. Tailored to a specific task, they allow developers to save time and programming effort, as satisfactory implementations of systems in popular programming languages can be quickly obtained, and applications that use filter banks can be rapidly prototyped without compromising much on efficiency.

Acknowledgments. This research work was supported by the European Union in the framework of the European Social Fund.

References

[1] Vaidyanathan, P.P.: Multirate systems and filter banks. Prentice-Hall, Englewood Cliffs (1993)

[2] Strang, G., Nguyen, T.Q.: Wavelets and filter banks. Wellesley-Cambridge Press, Wellesley (1996)

[3] Piotrowski, A., Parfieniuk, M.: Cyfrowe banki filtrow: analiza, synteza i implementacja dla systemow multimedialnych. Wydawnictwo Politechniki Bialostockiej, Bialystok, Poland (2006)

[4] Vaidyanathan, P.P.: On coefficient-quantization and computational roundoff effects in lossless multirate filter banks. IEEE Trans. Signal Process. 39, 1006–1008 (1991)

[5] Parfieniuk, M.: A dedicated high-level language for implementing nonrecursive filter banks and transforms. In: Proc 17th Int. Conf. Mixed Design of Integrated Circuits and Systems (MIXDES), Wroclaw, Poland (2010)

[6] Parfieniuk, M., Petrovsky, A.: Inherently lossless structures for eight and six-channel linear-phase paraunitary filter banks based on quaternion multipliers. Signal Process 90, 1755–1767 (2010)

[7] Parfieniuk, M., Petrovsky, A.: Quaternion multiplier inspired by the lifting implementation of plane rotations. IEEE Trans. Circuits Systems I 57, 2708–2717 (2010)

[8] Loeffler, C., Lightenberg, A., Moschytz, G.: Practical fast 1-D DCT algorithms with 11 multiplications. In: Proc IEEE Int. Conf. Acoust. Speech Signal Process. (ICASSP), Glasgow, Scotland (1989)

[9] Xiong, J., et al.: SPL: a language and compiler for DSP algorithms. ACM SIGPLAN Notices 36, 298–308 (2001)
[10] Püschel, M., et al.: SPIRAL: Code generation for DSP transforms. Proc. IEEE 93, 232–275 (2005)
[11] Locke, C.D., Dibble, P.C.: Java technology comes to real-time applications. Proc. IEEE 91, 1105–1113 (2003)
[12] Labrosse, J., et al.: Embedded software: know it all. Newnes/Elsevier, Oxford, UK (2008)
[13] Parr, T.: The definitive ANTLR reference: Building domain-specific languages. Pragmatic Bookshelf, Raleigh (2007), http://www.antlr.org

Singularities Detection System Design for Automatic Analysis of Biomedical Signals and Machine Condition Monitoring and Fault Diagnostics

Pawel Tadejko and Waldemar Rakowski

Abstract. Most important information about nature of a signal is often carried by the singularity points, such as the peaks, the discontinuities, etc. Moreover, at the moment when specific event occur, the output signals usually contain jump points that often are singularity points. Therefore, singularity detection has played an important role in signals processing, biomedical, e.g. ECG/EEG event detection, machine condition monitoring and fault diagnostics, etc. The wavelet modulus maxima method has been widely used method for the detection of singularity points. A review of the literature shows that the mentioned method may give very high efficiency in detection of events in the noisy ECG/EEG waveforms and also exist many application of the wavelet in machine fault diagnostics. Actually, both types of systems can be called CBMS - Condition Based Monitoring Systems. Because, we monitor the digital signal for search the specific events in both cases. The paper presents the proposal of system design for biomedical signal and vibration analysis and monitoring of machines, based on the Mallat and Hwang wavelet singularity analysis.

1 Introduction

The application of the wavelet transform for biomedical signal analysis and machine fault diagnostics has been developed for last 10 years at a very rapid rate.

Signal analysis is one of the most important methods used for automatic medical diagnostic and condition monitoring and fault diagnostics, whose aim is to find a simple and effective transform to the original signals. Therefore, the important information contained in the signals e.g. dominant features of signals can be extracted for diagnostics. Many signal analysis methods have been used for diagnostics, among which the FFT is one of the most widely used. Unfortunately, the

Pawel Tadejko · Waldemar Rakowski
Faculty of Computer Science, Bialystok University of Technology
Wiejska 45a, 15-307 Bialystok, Poland
e-mail: p.tadejko@pb.edu.pl, w.rakowski@pb.edu.pl

P. Lipiński, K. Świrski (Eds.): Towards Modern Collaborative Knowl. Sharing Syst., SCI 401, pp. 101–117.
springerlink.com

FFT-based methods are not suitable for non-stationary signal analysis and are not able to reveal the inherent information.

The wavelet transformation can automatically change the length of the window, and it can localize the signal in the space and the frequency domain and give the signal a partial characteristic. Analysis of the local signal properties is of great importance to the signal processing, because the so-called *singularity* very often carries information important for further processing.

Based on the singularity theory of wavelet transformation [12, 13, 14] we propose signal analysis system design for singularities detection in automatic analysis of biomedical signals and machine condition monitoring and fault diagnostics.

Results obtained in many papers on ECG signal analysis [1, 3, 9, 11, 16, 21, 25] presents that the proposed algorithm correctly detects the events, even under the presence of noise, baseline drift, muscle noise and artifacts due to electrode motion. On the other hand, there exist many application of the wavelet in machine condition monitoring or fault diagnostics [10, 19, 20, 24, 27] including the following aspects: the time-frequency analysis of signals, the fault feature extraction, the singularity detection for signals, the denoising and extraction of the weak signals, the compression of vibration signals and the system identification.

The aim of this paper is to analysis and design the software system for signal analysis and singularity detection. The concept of system implies the ability to analyze on-line and off-line data. Component for detection allows us to set parameters for monitoring condition and choose predefined signal type.

The key algorithm of the system involves the use of Mallat and Hwang wavelet singularity analysis. We propose an original, new approach to calculating of adaptive threshold that exploits statistical properties of the observed signal and additional heuristic. The threshold is independent for each successive signal window and the algorithm uses the properties of a series of distribution with a compartments class.

2 Wavelet Transform and Singularity Analysis

The fundamentals of singularity detection in signal using the wavelet transform have been shown by Stephane Mallat in [12, 13, 14].

Signal $f(t)$ smoothed by the function $\theta(t)$ can be represented as the result of the convolution

$$f(t) = f \star \overline{\theta}(u), \tag{1}$$

where

$$\overline{\theta}(t) = \theta(-t). \tag{2}$$

The derivative of the smoothed signal is equal to the convolution of the signal $f(t)$ with the derivative of smoothing function $\overline{\theta}'(t)$

$$\frac{d}{du}(f \star \overline{\theta})(u) = (f \star \overline{\theta}')(u) \tag{3}$$

where

$$\overline{\theta}'(t) = \frac{d\overline{\theta}(t)}{dt}. \tag{4}$$

The smoothing function and its argument can be scaled

$$\theta_s(t) = \frac{1}{\sqrt{s}}\theta\left(\frac{t}{s}\right), \tag{5}$$

where s is the factor of the scale. For $s>1$ the function $\theta(t)$ is dilated and average of $f(t)$ is performed on a wider range of the independent variable t.

The derivative of the smoothed signal is given by the convolution of the signal with the scaled derivative of the smoothing function

$$\frac{d}{du}(f \star \overline{\theta}_s)(u) = f \star \overline{\theta}_s'(u), \tag{6}$$

where

$$\overline{\theta}_s'(u) = \frac{d\overline{\theta}_s(u)}{du}. \tag{7}$$

In order to show the connection between the *Continuous Wavelet Transform* (CWT) and the derivative of the smoothed signal, we define wavelet $\varphi(t)$ in the form of the derivative smoothing function with the changed sign

$$\psi(t) = -\frac{d\theta(t)}{dt}. \tag{8}$$

The result is

$$\overline{\psi}(t) = \frac{d\overline{\theta}(t)}{dt} \tag{9}$$

and

$$\overline{\psi}\left(\frac{t}{s}\right) = s\frac{d}{dt}\left[\overline{\theta}\left(\frac{t}{s}\right)\right], \tag{10}$$

which means that

$$\overline{\psi}_s(t) = s\overline{\theta}_s'(t). \tag{11}$$

The CWT is determined and defined as follows [15]

$$Wf(u,s) = f \star \overline{\psi}_s(u). \tag{12}$$

By substituting in (12) with $\overline{\psi}_s(t)$ the right side of (11) θ, we get

$$Wf(u,s) = s\ (f \star \overline{\theta}_s')(u). \tag{13}$$

Based on the above equations and (6), we get

$$Wf(u,s) = s\frac{d}{du}\left(f \star \overline{\theta}_s\right)(u). \tag{14}$$

To differentiate the signal smoothing by the function $\theta(t)$, it is enough to calculate the CWT of the signal with the wavelet defined by the equation (8).

The domain of CWT is plane *0us*. For calculation purposes, it is necessary to discretize variables u and s. In the following, it is assumed that the signal $f(t)$ is analyzed in the range $[0,N-1]$.

2.1 Dyadic Wavelet Representation of a Signal

For the purpose of singularity analysis, a *dyadic wavelet transform* of the signal $f(t)$ can be used for with the scale s takes dyadic values, i.e. $s=2^k$, $k \in Z$.

The digital signal

$$a_0 = (a_0[n])_{n\in Z}, \tag{15}$$

should be calculated by the formula

$$a_0[n] = \int_{-\infty}^{+\infty} f(t)\phi(t-n)dt, \quad \mathrm{n} \in \mathrm{Z}. \tag{16}$$

where $\phi(t)$ is the scaling function of the wavelet analysis.

Scaling ϕ by dyadic step we can get a series of digital representation

$$(a_k)_{k\in Z}, \tag{17}$$

where

$$a_k[n] = \int_{-\infty}^{+\infty} f(t)\frac{1}{\sqrt{2^k}}\phi\left(\frac{t-n}{2^k}\right)dt, \quad \mathrm{n} \in \mathrm{Z}. \tag{18}$$

Averaging of the signal is performed at a distance proportional to the scale 2^k.

Detailed signals d_k are calculated by analogous to the signals a_k.

$$(d_k)_{k\in Z}, \tag{19}$$

where

$$d_k[n] = \int_{-\infty}^{+\infty} f(t)\frac{1}{\sqrt{2^k}}\psi\left(\frac{t-n}{2^k}\right)dt, \quad \mathrm{n} \in \mathrm{Z}. \tag{20}$$

Starting from the digital signal a_0 given by (16), we calculate the *dyadic discrete wavelet transform* of a signal as:

$$\{(d_k)_{1\leq k\leq K}, a_K\}, \tag{21}$$

where a_K is a coarse representation of the signal at scale 2^K, and d_k is a detail signal at scale 2^k, wherein $1\leq k\leq K$. Most of the details are in the signal d_1, least in d_K.

For the purposes of the singular points detection in a signal, it is sufficient to perform only signal analysis, i.e. to find the dyadic discrete wavelet transform. If the coefficients $a_0[n]$ are zero for $n<0$ and $n\geq N$ then the dyadic discrete wavelet transform corresponds to the grid points of K rows, $K\leq log_2 N$ for N points in each row. Detection of singular points is associated with searching modulus maxima of wavelet coefficients *centered* around the vertical lines of the grid corresponding to fixed values of u.

2.2 The à trous Algorithm of Dyadic Wavelet Decomposition

Coefficients defined by (18) and (20) can be calculated iteratively using the so-called *à trous algorithm* proposed by Holschneidera, Kronland-Martineta, Morleta and Tchamitchiana [8, 23].

After selecting the scaling function ϕ and wavelet φ based on the wavelet equations there calculated two digital filters:

the low pass filter h and high pass filter g.

Let, h_k is filter obtained by inserting 2^k-1 zeros between each pair of impulse response filter coefficients h. The extension of the filter by inserting zeros creates *holes* (franc. trous) that are mentioned in the name of algorithm. By definition

$$\overline{h}_k[n] = h_k[-n]. \tag{22}$$

The same designations apply to the filters g and g_k.

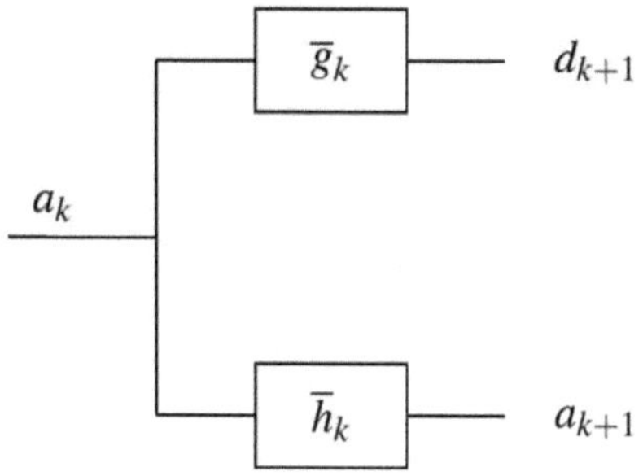

Fig. 1 Filter bank implementing one step of *à trous* algorithm.

Coefficients of the dyadic discrete wavelet transform of a signal a_0 can be calculated using the following iterative calculation scheme [15]

$$d_{k+1}[n] = a_k \star \overline{g}_k[n], \tag{23}$$

$$a_{k+1}[n] = a_k \star \overline{h}_k[n] \tag{24}$$

for $k=0,1,2,\ldots$

If the input signal $(a_0[n])_{n\in Z}$ has a finite length of N samples, then the operations (23) and (24) can be achieved by using circular convolution (recurring).

2.3 Singularity Detection Based on Wavelet Multiscale Analysis

If we record physical phenomena as a digital signal - it can be the basis for the analysis of singularities in such a signal. Currently, they are increasingly used short-time and wavelet analysis, which enable more efficient extraction of the singular characteristics of short-analyzed signals. These indicators may be related to the occurrence of specific phenomena or damage.

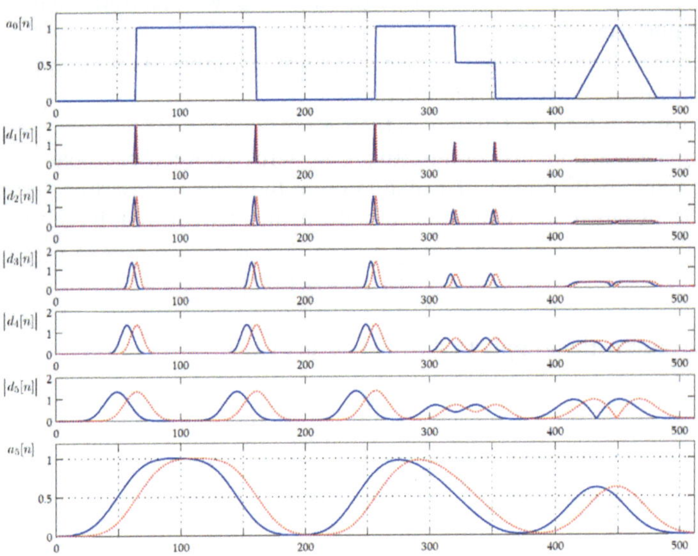

Fig. 2 Modulus maxima of approx $a_k[n]$, $(k=0)$ and details $d_k[n]$, $(k=1..5)$ coefficients calculated by iterated version of à trous algorithm for some singularities in the signal.

As we mentioned, many scientist use wavelet singularity analysis for ECG signal processing [1, 3, 9, 11, 16, 21, 25, 26] or for machine condition monitoring or fault diagnostics [10, 19, 20, 24, 27].

Therein, the correspondence between singularities of a function $f(t)$ and local maxima in its wavelet transform $Wf(u,s)$ is investigated. The authors of these works have used the dyadic wavelet transform (DyWT) for detecting singular points. It is shown that singularities correspond to pairs of modulus maxima across several scales (Fig. 2). The figure clarifies the correspondence between a signal with singularities and its wavelet coefficients.

Analysis of singularities characterizing biomedical signals and measurement concerns the local modulus maxima of dyadic wavelet transform. Singularity points, such as the QRS complex, the peaks, the discontinuities, etc. of the signal are represented by local maxima at successive levels of decomposition. Location of the singularities is dependent on the singularities. Characteristic points are detected by comparing the coefficients of the discrete DyWT of selected scales against fixed thresholds. R-peaks are detected when the locations of modulus maxima of adjacent scales exceed a threshold that is calculated for every segment.

Digital filters of wavelet analysis corresponds a quadratic spline wavelet and a quadratic box spline scaling function. This means that in signal analysis and detection of singularities, we can precisely control the parameters of the separation process for selected frequencies.

2.4 An Adaptive Threshold for Events Detection

The great variability of patterns depends on the specific characteristics of the signals and change over time. The idea of a threshold for the detection algorithm presented in the paper uses the distribution of class intervals.

The adaptive method of threshold detection that we present in this paper is independent for each successive signal window. The threshold algorithm uses the properties of distribution with a compartments class.

Let the modulus maxima DyWT values of the processed signal window

$$x_{k,1},\ldots,x_{k,n_k} \tag{25}$$

will be n_k-element samples which contain the absolute modulus maxima values at each level k dyadic wavelet transform.

The *distance* of the *feature* X is the difference

$$R_k = x_{k,max} - x_{k,min}, \tag{26}$$

where $x_{k,max}$ and $x_{k,min}$ denotes the highest and lowest value in a sample.

The distance is thus the length of the shortest interval in which all values are within a sample. With a larger sample size, in which facilitates analysis, aggregated in *classes*. For simplicity, we assume that intervals is equal length. Let suppose that all values in a given class are identical to the measure class. There are several rules for determining the number of classes c_k depending on the cardinality n_k of the sample. In our research here, they are:

$$c_k \approx \sqrt{n_k} \quad or \quad c_k \approx 1 + 3.322 \log n_k, \tag{27}$$

If R_k is the range of the sample, the number of classes c_k, then b_k denotes the *length of class*, and we assume that $b_k \approx R_k/c_k$. The number of samples with values contained in the i-class is called cardinality (the size) i of the class, and we denote $n_{k,i}$ and take into account the k levels of dyadic wavelet decomposition $n_{k,i}$. The symbol $\overline{x}_{k,i}$ means the center of a subsequent class, whose values are mean values lower and upper limit for each interval i class. The result is a series of pairs of numbers: center of the class $\overline{x}_{k,i}$ and the cardinality $n_{k,i}$, called distribution with a compartments class.

Each window contains modulus maxima values of DyWT to be analysed. The value of the threshold for every DyWT scale is calculated as follows:

- two classes are calculated that contains the local maxima with the largest values - $x_{k,c_{k-1}} = x_{k,max} - 3/2 \cdot R_k / c_k$ and $x_{k,c_k} = x_{k,max} - 1/2 \cdot R_k / c_k$, for the distribution $\overline{x}_{k,i}$ values of modulus maxima for all scales of decomposition k.
- thresholds for QRS detection are established at each level k as central values in the interval $[x_{k,c_{k-1}}, x_{k,c_k}]$ for all decomposition scales k.

The occurrence of a events is detected by comparing the QRS candidates for selected level of all scales of DyWT. If the locations of the local maxima exceed the threshold correlated across two or tree consecutive scales, (Fig. 2), we assume that the locations of these maxima correspond to the location of singularity points, such as the peaks, the discontinuities, etc.

3 Application of Wavelet Singularity Analysis

While the wavelet singularity analysis was used to detect QRS complexes in ECG signal for many years [11], using to analysis and diagnostics is the subject of a fairly recent, e.g. engines or wind turbines [10, 19, 20, 27] and distortion of the power industry networks [24].

3.1 Biomedical Signals Analysis

All of the wavelet-based peak detection methods mentioned in this paper [1, 3, 9], 11, 16, 21, 25, 26] are based on Mallat and Hwang's approach for singularity detection [13].

In our researches, we propose a QRS complex detector based on the Mallat and Hwang singularity analysis algorithm which uses dyadic wavelet transform. We design a spline wavelet that is suitable for QRS detection. The scales of this decomposition are chosen based on the spectral characteristics of electrocardiogram records. By proceeding with the multiscale analysis we can find the location of a rapid change of a signal, and hence the location of the QRS complex. The performance of the algorithm was tested using the records of the MIT-BIH

Arrhythmia Database. The method is less sensitive to time-varying QRS complex morphology, minimizes the problems associated with baseline drift, motion artifacts and muscular noise, and allows R waves to be differentiated from large T and P waves.

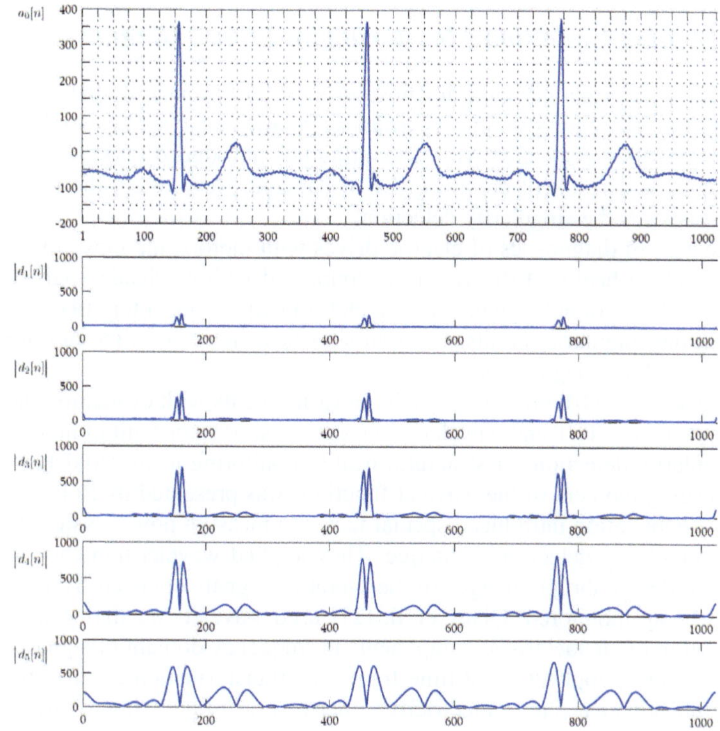

Fig. 3 The singularities in the ECG signal and its dyadic WT calculated by the *à trous* algorithm; $a_0[n]$- ECG signal, $|d_1[n]|$-$|d_5[n]|$- the modulus maxima of details coefficients in different scales.

The QRS complex and other singularities in the signal are represented by local maxima at successive levels of decomposition. R-peaks are detected when the locations of modulus maxima of adjacent scales exceed a threshold that is calculated for every segment. For most R waves, their energies are concentrated at scales $|d_3[n]|$ and $|d_4[n]|$.

The developed QRS detector based on wavelet singularity analysis of a signal consists of three main blocks:

- filtering ECG signal by a filter bank of dyadic wavelet transform (DyWT) using a quadratic spline wavelet and a quadratic box spline scaling function,
- detection of candidates for QRS complexes using locations of modulus maxima on selected scales with adaptive thresholds,

- determining the final list of QRS complexes using heuristic methods which use additional decision rules for reducing the number of false-positive detections.

The main part of the detection stage of the QRS complex is an original adaptive algorithm for the calculation of a QRS detection threshold.

3.2 Machine Condition Monitoring and Fault Diagnostics

Condition monitoring and fault diagnostics is useful for ensuring the safe running of machines and other systems. Wavelet singularity analysis takes advantage of the non-linearities associated with discontinuities introduced into the dynamic response data as a result of certain types of damage. Very often, the key issue in fault diagnosis is the singularity detection.

In the case of diagnostics of electric drives with motors, the most important information about health of the system is connected with mechanical vibration and its trace in the recorded signal. The wavelet transforms are often used for the detection of the singularity points in output signals sampled from the machines, furthermore, for fault diagnostics.

Kowalski et al. [10] presented results of experiments with evaluation the efficacy analysis FFT, STFT and wavelet in the detection of rotor fault in induction motor. Singularity detection for structural health monitoring using Holder exponents (the vanishing moment of the wavelet function) was presented by Robertson et al. [20]. The publication introduced special factor to measure power system fault, using the singularity detection technique. They applied wavelet transform modulus maxima to detect abrupt changes in the vibration signals obtained from operating bearings being monitored. Qiang et al. [27] used wavelet singularity analysis for identification of characteristic components in frequency domain of signal. Skomudek [24] showed application of time-frequency transformation for the analysis of wave phenomena in electricity networks. For improving power system efficiency, it is necessary to monitor the power quality disturbances, to acquire the node voltage and current, and locate the sources of power network disturbances.

Peng et al. [19] made a extensive review to present a summary about the application of the wavelet in machine fault diagnostics, including some aspects: the time–frequency analysis of signals, the fault feature extraction, the singularity detection for signals, the denoising and extraction of the weak signals, the compression of vibration signals and the system identification. Some other applications are introduced briefly as well, such as the wavelet networks, the wavelet-based frequency response function, etc. In addition, some problems in using the wavelet for machine fault diagnostics are analyzed. The prospects of the wavelet analysis in solving non-linear problems are discussed.

Using the time-frequency methods can extract these signals symptoms of damage in short and long periods of time (information high- and low-frequency). The majority of structural health monitoring studies [19] reported in the technical literature focus on detecting damage.

The wavelet transforms are very successful in singularity detection, but when using the wavelet transforms to detect singularity, regularity could be an important criterion in the selection of a wavelet function. We used quadratic spline wavelet and a quadratic box spline scaling function. In rotating machinery condition monitoring, identification of characteristic components is fundamental in many engineering applications so as to obtain fault sensitive features for fault detection and diagnosis.

3.3 Areas of Possible Applications of Solution Based on Singularity Detection System

Developed platform for analysis of signals can be used as a subsystem in monitoring, an advanced diagnosis and alarming system. System for automatic biomedical signal analysis can use the features detection function to detect life-threatening arrhythmias. Online condition monitoring systems can use singularity detection function for performing predictive maintenance.

Module detection and analysis of singularities in the digital signal is a tool supporting scientific research, conduct engineering work and didactics, as well as a tool for use by other users as a potential element of commercial solutions. The component can be used as part of the commercial systems in the following areas:

- analysis of singularities in biomedical signals – examples of commercial solutions [5, 17]
- signal processing for monitoring and diagnostic machine - examples of commercial solutions [2, 4, 7, 18, 22],
- the analysis of measurement data signal, such as power industry networks – examples of commercial solutions [4, 6].

Services system mentioned above can be used in such areas as the following functionality:

- Analysis of singularities in biomedical signals, such as analysis of heart rate, detection heart beats (QRS complexes in ECG signal). QRS detector is the most basic element of ECG monitoring and analysis e.g. with Holter method.
- Signal processing for monitoring and diagnostics of machines and electric drives. Most commonly used in fault detection of rotating elements - by detecting the specific frequency of vibration.
- Analysis of measurement data signal, such as the power network - pulse distortion, surges, etc. e.g. detection of faults in the systems of power station.

Monitoring systems should provide not only functionality such as alarm and event handling, status monitoring, but also advanced functions like disturbance data analysis and evaluation, primary apparatus monitoring, power quality monitoring, and condition based maintenance. The monitoring system should be open, i.e. different data sources, devices and software components should be easily integrated into the system; legacy systems can be replaced by new components without much effort.

4 The Software Architecture of a System Services for Singularity Detection

The component of designed systems consists of sensing, monitoring, analyzing, disease or fault detection, and emergency alert. In our system, services for singularity detection are responsible for event monitoring with particular attention to detection specific symptoms. The system also provides an application for import recording activities, events and other diagnostics signals.

We choose the Rich Internet Application (RIA) architecture because represents the evolutionary user interface is web browser-independent. Once the server responds, the browser will parse the server's results and decide the best way and the appropriate time to show the results to you. This is how the original client-server applications functioned. These thick-clients had the drawback that they had to be deployed to the client machine they ran on, which was often a manual and error-prone process.

Browsers eliminated these issues, but browsers also lacked the rich capabilities of their client-sever predecessors, capabilities enterprise applications require. RIA's goal is to be the best of both worlds. It aims to achieve the richness of client-server functionality with the ease of browser-based deployment.

We choose Adobe because Flex is a mature technology and has the most mature component library for RIA development. That makes the use of Flex as front-end and Java as back-end server side technology for developing RIA applications.

The overview of the system communication is shown in Fig. 4. As can be seen in Fig. 4, is a distributed system, i.e. it consists of a number of subsystems and components.

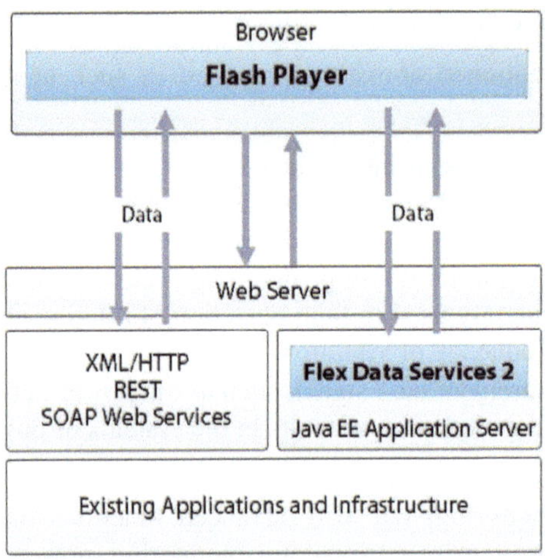

Fig. 4 Architecture of a Flex application.

With evaluation and analysis programs installed it evaluates data/information from different sources through data interfaces. Usually it is located remote to field devices. Subsystem is a part system for the tasks in one substation: data acquisition, data storage, data visualization, data evaluation and monitoring.

In service analysis sample data is introspected after it is obtained from one or more providers that will be queried by the application. If the speed of the network is enough, the software may gather data online per-second with network communication and display real-time monitor function on the interface. The system may gather data in a synchronous whole-cycle approach way and gather monitoring information when external system works or in an asynchronous mode.

5 Requirements Analysis and Conceptual System Design

Implementation of the monitoring system measurement, due to the need for data acquisition and test and analysis, So it need to adopt computer technology to automatically detect the real time data processing, curve drawing, precision analysis, etc. This module is the most critical section of the system and its main functions are:

- supervising the overall operation of the system
- collecting data from the input modules after occurrence of a fault and organizing the acquired data in a file
- post processing of the data for calculating the virtual and logical channels
- sending fault information in the different formats for using in analyzer tools
- managing mass storage of the subsystem
- the possibility of a block detector design by setting parameters detection
- the possibility to use library of test signals and the loading their own samples,
- the ability to interact with the user through a GUI
- the possibility of a visual analysis of results and test parameters influence the results of the analysis of singularities.

It is obvious that all these tasks must be done properly and simultaneously. So a multi-task programming library has been chosen for the system and software of this module to implementation.

In order to determine the functional requirements of the proposed system we defined users: lecturer, scientist, engineer, and medical doctor, represented in the modeling process by the so-called UML - actors.

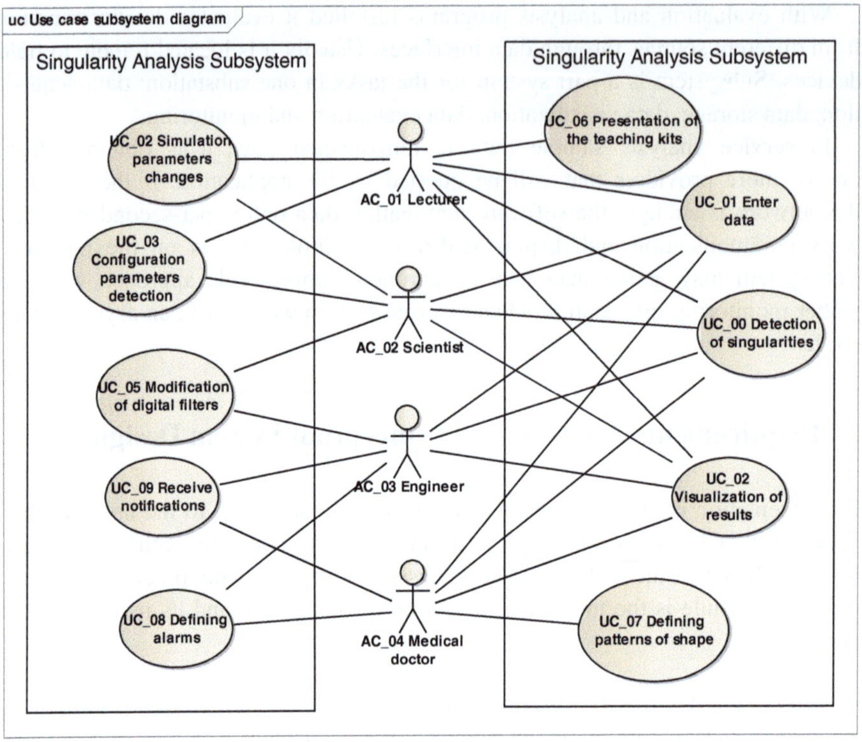

Fig. 5 Use case diagram of designed singularity detection subsystem.

Some use cases is the core of the system and is shared by all actors. These are the main event of the detection algorithm, data management and visualization. Others are specific to certain actors, and they are about their separation.

For a fuller description of the requirements we plan drawing up the so-called use case scenarios. They are presented in text form and determine the sequence of activities performed by the actor or the system. They will set input conditions for initiating a case, the system state is characterized by the implementation of the case and the main scenario. The defined list of possible alternative scenarios and non-functional requirements relevant to the implementation (implementation of the subsystem) use case.

The prospect of a conceptual system are modeled using class diagrams. Objects have been identified relevant to achieving the objectives of the system. Basic classes have been identified which are a generalization of existing facilities. We are working now on detailed analysis, which will define the methods highlighted classes (sets of operations) implemented in their subsystem.

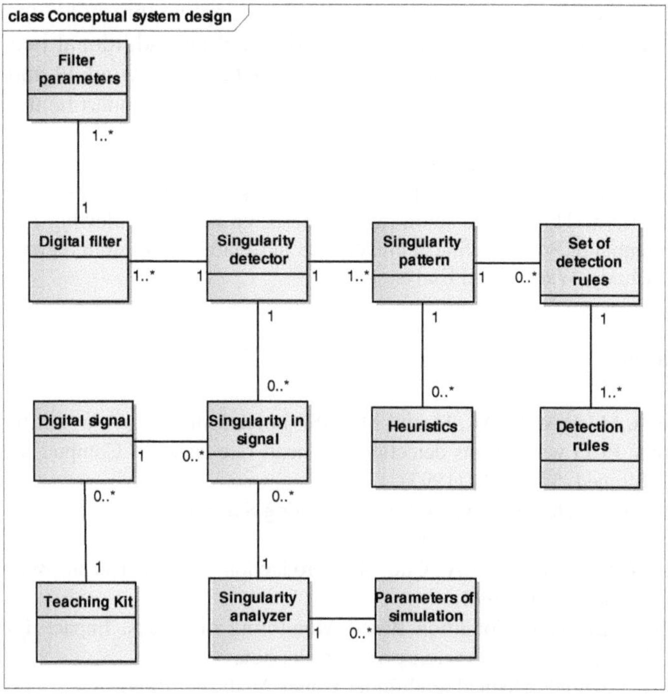

Fig. 6 Conceptual model of the singularities detection and analysis system. Chart includes only the major classes, which are the core concepts.

6 Conclusions

This paper shows how a component based approach is used for singularity detection system design. The input of the approach is based on the requirements specification of the automatic analysis of biomedical signals and machine condition monitoring and fault diagnostics. The output is a possible solution architecture of the system.

This paper illustrates that the application of the object oriented methodology UML technology are feasible to reach the goals. In the following part of this report, overviews of UML subsystem are given in section 4. The design of our singularity detection is presented from a static structural point of view, i.e. the Use Case diagram and the Class diagrams are presented and analyzed.

However, given the skeleton of the system, these additional functional modules can be added to both the static and the dynamic descriptions of the system, in the future. There are some aspects of singularity detection subsystem that need to be addressed:

- definition of task and task communication;
- time constraints;
- modeling of the network.

We propose an original, new approach to define adaptive threshold algorithm that exploits statistical properties of the observed signal and additional heuristic. The threshold is independent for each successive ECG signal window and the algorithm uses the properties of a series of distribution with a compartments class.

Acknowledgments. This paper was partially supported by grant of Faculty of Computer Science, Bialystok University of Technology, Bialystok, no. S/WI/4/08 and partially by Information Platform TEWI - the supra-regional research network, allowing for integration scientific potential of Polish technical universities in Central and Eastern regions: Technical University of Lodz, Warsaw and Bialystok.

References

[1] Bahoura, M., Hassani, M., Hubin, M.: DSP implementation of wavelet transform for real time ECG wave forms detection and heart rate analysis. Comput. Methods Programs Biomed. 52, 35–44 (1997)
[2] Nevada, B.: Trendmaster Condition Monitoring System (2010), `http://www.bently.com`
[3] Burrus, C.S., Gopinath, R.A., Guo, H.: Introduction to Wavevlets and Wavelet Transforms. Prentice Hall (1998)
[4] CBMi - Real Time Condition Based Monitoring of Assets. Impact Technologies, LLC (2009), `http://www.impact-tek.com`
[5] Del Mar Reynolds Medical. Pathfinder Holter Analyzer (2009), `http://www.spacelabshealthcare.com`
[6] GE Energy, Electrical Equipment, Condition Monitoring and Diagnostics (2010), `http://www.gepower.com/system1`
[7] GE Energy, Wind Turbine Condition Monitoring and Diagnostics (2010), `http://www.ge-energy.com/wind`
[8] Holschneider, M., Kronland-Martinet, R., Morlet, J., Tchamitchian, P.: A Real-Time Algorithm for Signal Analysis with the Help of the Wavelet Transform. In: Wavelets, Time-Frequency Methods and Phase Space, pp. 289–297. Springer, Heidelberg (1989)
[9] Kadambe, S., Murray, R., Boudreaux-Bartels, G.F.: Wavelet Transform - based QRS complex detector. IEEE Transaction on Biomedical Engineering 46, 838–848 (1999)
[10] Kowalski, C.T., Kanior, W.: Evaluating the efficacy analysis FFT, STFT and wavelet in the detection of rotor fault in induction motor. In: Scientific Papers of the Institute of Electrical Machines, Drives and Measurements, vol. 60, Wroclaw University of Technology (2007)
[11] Li, C., Zheng, C., Tai, C.: Detection of ECG characteristic points using wavelet transforms. IEEE Transaction of Biomedical Engineering 42, 21–28 (1995)
[12] Mallat, S.: Zero-crossings of a wavelet transform. IEEE Transactions on Information Theory 37(4), 1019–1033 (1991)
[13] Mallat, S., Hwang, W.L.: Singularity detection and processing with wavelets. IEEE Transaction on Information Theory 38, 617–643 (1992)
[14] Mallat, S., Zhong, S.: Characterization of signals from multiscale edges. IEEE Transactions on Pattern Analysis and Machine Intelligence 14(7), 710–732 (1992)
[15] Mallat, S.: A Wavelet Tour of Signal Processing, 2nd edn. Academic Press (1999)

[16] Martinez, J.P., Almeida, R., Olmos, S., Rocha, A.P., Laguna, P.: A wavelet-based ECG delineator: evaluation on standard databases. IEEE Transactions on Biomedical Engineering 51, 570–581 (2004)
[17] Mortara Instrument. Diagnostic Cardiology Company WI and Los Altos, CA (2009), http://www.mortara.com
[18] MTI Instruments. Vibration and Balancing System (2010), http://mtiinstruments.com
[19] Peng, Z.K., Chu, F.L.: Application of the wavelet transform in machine condition monitoring and fault diagnostics: a review with bibliography. Mechanical Systems and Signal Processing 18, 199–221 (2004)
[20] Robertson, A.N., Farrar, C.R., Sohn, H.: Singularity detection for structural health monitoring using holder exponents. Mechanical Systems and Signal Processing 17, 1163–1184 (2003)
[21] Sahambi, J.S., Tandon, S.N., Bhatt, R.K.P.: Using wavelet transforms for ECG characterization. An on-line digital signal processing system. IEEE Engineering in Medicine and Biology Magazine 16, 77–83 (1997)
[22] Sensor Highway. Condition Based Monitoring Systems, Vibra Metrics (2010), http://www.vibrametrics.com/
[23] Shensa, M.J.: The Discrete Wavelet Transform: Wedding the à Trous and Mallat Algorithms. IEEE Transactions on Signal Processing 40(10), 2464–2482 (1992)
[24] Skomudek, W.: Application of time-frequency transformation for the analysis of wave phenomena in electricity networks. Polich Energetic Networks Operator SA (2007)
[25] Strang, G., Nguyen, T.: Wavelets and Filter Banks. Wellesley-Cambridge Press (1996)
[26] Szilagyi, L., Benyo, Z., Szilagyi, S.M., Szlavecz, A., Nagy, L.: On-line QRS complex detection using wavelet filtering. In: Proceedings of the 23rd Annual International Conference of the IEEE Engineering in Medicine and Biology Society, vol. 2, pp. 1872–1874 (2001)
[27] Qiang, M., Dong, W., Hong-Zhong, H.: Identification of characteristic components in frequency domain from signal singularities. Review of Scientific Instruments 81, 035113–035113-7 (2010)

Proposal of New Systolic Architecture for Mathematical Morphology Algorithms

Krzysztof Lichy

Abstract. This paper is devoted to the problem of designing an optimal systolic array for morphological processing. The paper gives the proposal of the new systolic architecture with the examples of using new features.

1 Introduction

Undertaking the designing of the systolic array architecture it is important to take into consideration a very important issue, which influences all the further process of designing. The question whether the designed array is to have universal architecture or whether it is designed to a specific purpose is to be answered. If the answer is positive to the second part of the above statement the solution is obvious – the array must be as compatible to the given problem as possible. Otherwise the architecture should be as universal as possible in order to do the best matching to a lot of mathematical problems. The example of such a solution is the systolic array Systola 1024 (made by ISATEC, its description is too extended to be presented here, that is why it can be found in [6],[9]).

Dedicated solution is less complicated thus cheaper. Universal solution because of its complexity is usually more expensive but it can be used to a lot of different tasks, which makes it more flexible. A certain analogy to the issue occurring in the 1980s can be found. What is better, a cheaper automatic machine, which can do limited chain of movements or the limitless more expensive robot. [12],[15]

Unfortunately in the case of systolic array such an attitude to the architecture designing does not turn out to be practical. Thus, it is important to tend connecting both ideas. In the article the proposal of the new systolic array architecture will be presented.

2 The Author's Universal Array Project

In the presented project according to the idea of the systolic architecture, the array consists of identical simple processing elements (PE) connected only to the closest

---

Krzysztof Lichy
Technical University of Łódź, Institute of Information Technology
Wolczanska 115, 90-924 Lodz, Poland
e-mail: krzysztof.lichy@p.lodz.pl

P. Lipiński, K. Świrski (Eds.): Towards Modern Collaborative Knowl. Sharing Syst., SCI 401, pp. 119–132.
springerlink.com

neighbors. Each PE has included 32 registers as it is in Systola 1024. Each PE can do logical operation and simple math operation (such as +,-,*,/,^) on the data situated in the registers. Moreover, as the array is designed to mathematical morphology processing, each PE has implemented function min and max operating on two registers and showing higher or lower value. Such a set of operation will make it possible to undertake every morphological operation and it will make possible to solve most math problems. Basing on the ISATEC solution the array consists of 32x32 PE connected orthogonal.

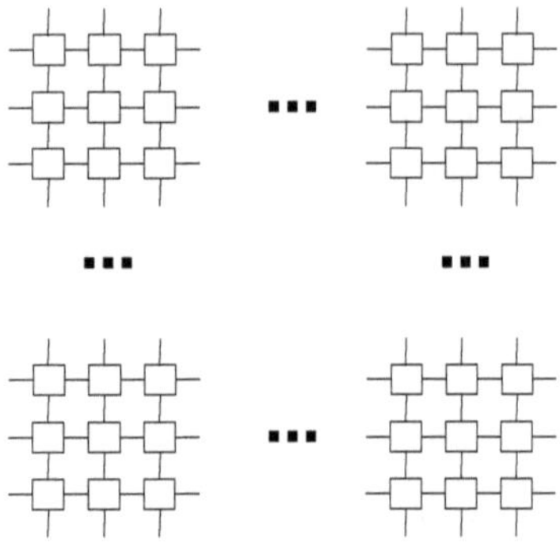

Fig. 1 Systolic array 32x32 size [10]

When compared to Systola 1024 a certain modification of loading data have been introduced. Like Kung and Leiserson [2] solution data to the array can be moved in and out from each of the directions (described an N-North, E – east, W – West, S – South).

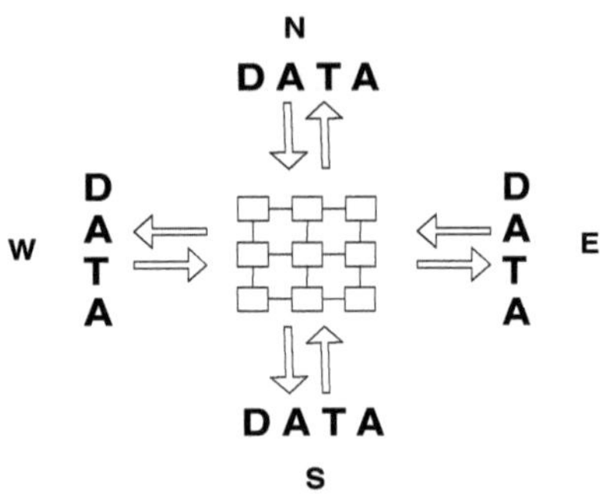

Fig. 2 Possible directions of floating data [10]

Because not all of algorithms can be easily implemented to the 32x32 array, there is a possibility of connecting arrays in bigger regular structures. An example of this problem is making an erosion operation on the picture that is bigger than the size of the systolic array. Instead of designing the program, which can make division and compact of the picture, the size of the systolic array is done by connecting smaller arrays.

As mentioned earlier, each PE has 32 computational registers and 32 communication registers. Each register is 32bit size, this is enough for the most of the mathematical problems. Bigger size of the registers seems to be overflow solution. The names of registers are analogical to ISATEC firm solution (the more detail description can be found in [9]). It is important to notice that in Systola 1024 there is only one communication register. Registers marked as a special ones are the communication registers C1..16 (size 32 bits). In the systolic array the register C1..16 of each PE situated to the north is seen as its own register CN1..16 analogical in different directions for example CS1..16.

A certain novelty, as compared to Systola 1024, are special communication registers SC1..16 that feature twice longer time of sending data to the neighboring PE (two pulsations instead of one) which is important facilitation with some systolic algorithms.

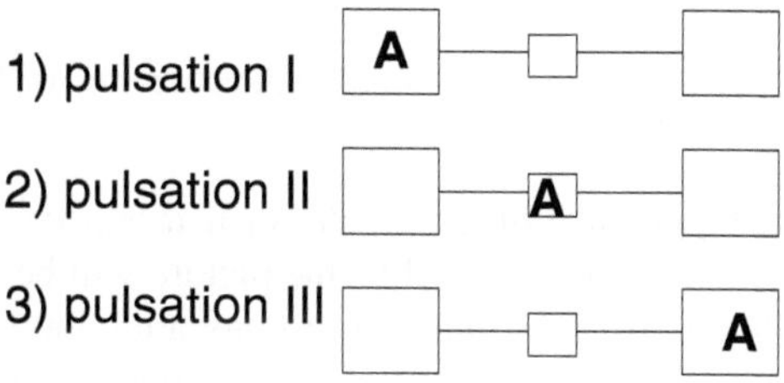

Fig. 3 Special communication register [10]

3 Introducing the Picture to the Systolic

First, the pixels of the picture to the particular PE should be introduced. This way particular pixels of the picture are situated in an appropriate PE.

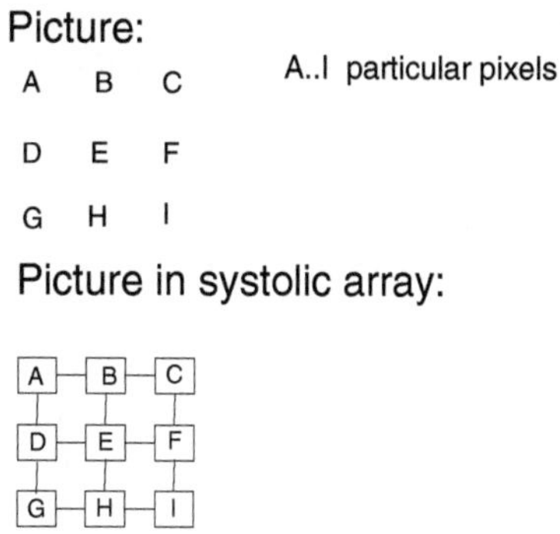

Fig. 4 Introducing picture to systolic array [10]

If the picture is of a small size, for the simplification, an introduction from one direction can be assumed.

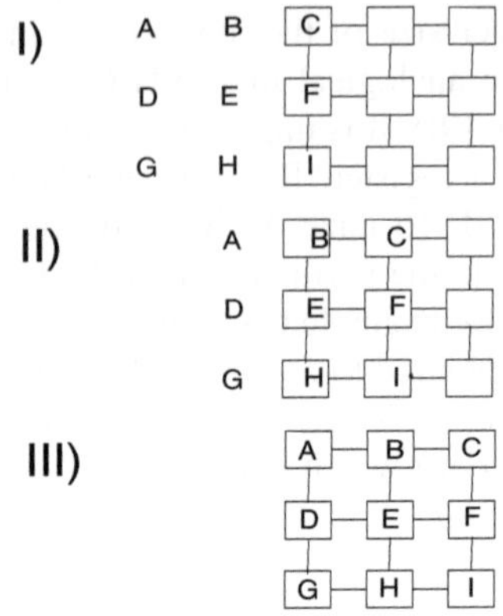

Fig. 5 Introducing picture to systolic array from one direction (west) [10]

In the case of the bigger size of the picture or if it is necessary to do the morphological operation as quickly as possible, the picture will be introduced simultaneously from two directions. It is important to pay attention to the problem with the proper division of the picture into the parts introduced from two directions. It may turn out that shortening the time of loading the picture to the array twice will demand too much effort when preparing the picture for loading. This problem is not important when the array is used only for morphological operation and all introductonary treatment of the picture is done by another specialized system.

When analyzing the issue of introducing the picture to the systolic array it should be noticed that this operation is the most time consuming in the whole process of image processing. For example for the 4x4 size array and for picture 4x4 pixels size with the above assumption the introduction of the picture will take the time of two pulsations (without the time necessary for proper division of the picture). Erosion operation done with the help of the typical algorithm will take time at least twice longer (four or more pulsations). After the erosion operation loading out the results will take two pulsations as well. Thus, it should be noticed that introducing and loading out the data will consume the time equal or little shorter to the time of doing a morphological operation without taking into consideration time of division and assembling of the picture. It is the feature to be remembered when creating the systolic algorithms of the picture processing.

There is a question why the data is to be introduced and loaded out only from two directions if the array is adjusted to exchanging data from all the directions. Unfortunately, introducing data from all the directions will not change the time necessary for placing the picture in the array.

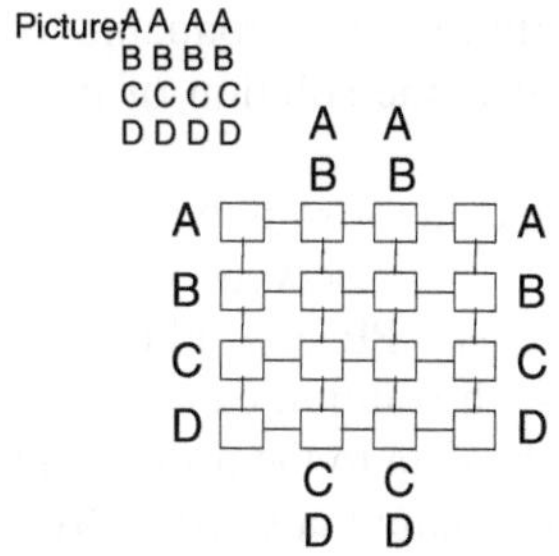

Fig. 6 Problem with introducing picture from all directions [10]

Such a solution will not accelerate the time of introducing. Moreover it may lead to increase in the complication with the preparation of the picture.

4 Possible Modification of the Systolic Array

Two-dimensional construction of the systolic array in general excludes the possibility of considerable acceleration of introduction and leading the data out from the array. Thus, introducing two special (32 bits size) communication registers to each PE, which are only for immediate communication with the communication register outside array can be proposed.

The solution is complianced with the definition of systolic array proposed by Kung because it is only the modification relying on the introducing the immediate possibility of communication with each processor. Two of such registers may give the possibility of creating three-dimensional systolic array with another direction of data flowing in the future. This solution would make sequential making of the morphological filters on the picture a very quick process. For example the open operation, which consists of sequentional erosion operation and dilatation operation. [18][19]

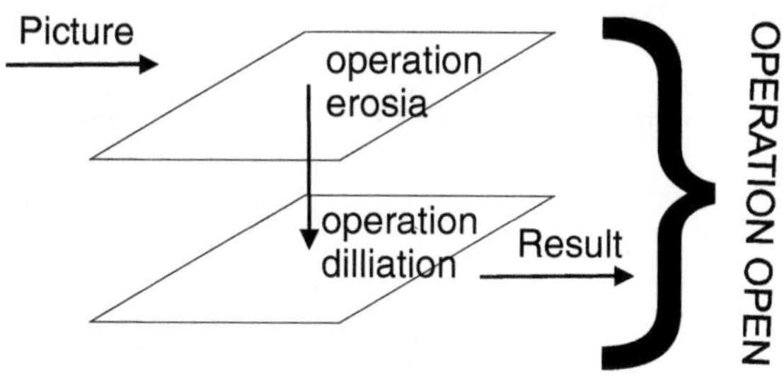

Fig. 7 3-dimensional systolic array [10]

Because there is a possibility of introducing/leading out data to/from the IN/OUT register of any PE it is the solution of problem introducing/leading out date to/from array.

Introducing/leading out the data to/from the array in this method is optional and it is why reverse compability is preserved. The disadvantages of such IN/OUT register solution is making every PE and addressing particular PE much more complicated. Theoretically, such solution can undergo a little modification which simplifies addressing a lot. The data to the array, which is situated on the "top" of the stack, is applicated to the method applied in Systola 1024 (from one ore two directions). The data to the array situated below is transmitted by the IN/OUT register. Whereas the data from the array from the "bottom" of the stack is led out with the method applied in Systola 1024. In such the solution the conception of addressing the array is not modified and the array is perceived as one. In the case of simple calculations the arrays placed below will not be used. Moreover, as previously reverse compability will be preserved.

5 Further Proposals of the Array Modification

It should be noticed that for structural element with the below shape (Fig. 8.) the orthogonal architecture is an optimal solution. Nevertheless, for the popularly occurring shape of structural element (3x3 size) because of the shape of the array, algorithms need to take into the account the necessity of communications with PE situated aslant to the PE consisting the origin. In order to avoid such complication additional connections between PE based on well known hexagonal architecture should be introduced. It is an improvement consisting of adding to already proposed architecture connections between PE situated diagonally below and under provided PE.

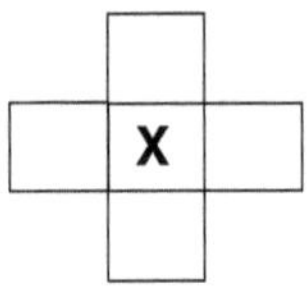

Fig. 8 Structural element (x – origin) [10]

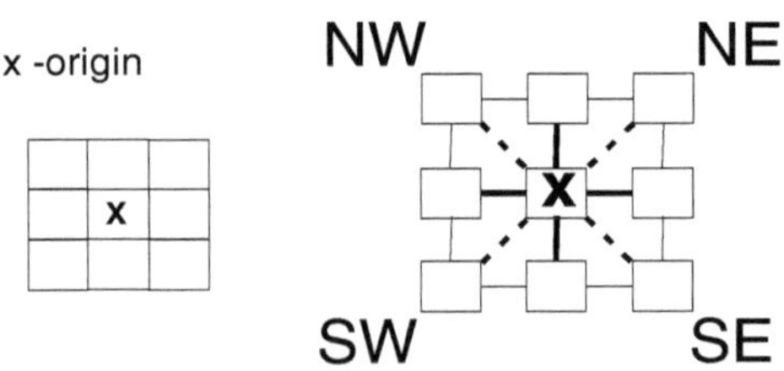

Fig. 9 New connection between PE [10]

In the above array communication register C of PE situated diagonally is analogical to communication registers described earlier (directions NW, SW, NE, SE).

6 Another Approach to Introducing Data

As mentioned earlier the operation of introducing the picture to the array is the most time consuming element of its processing. It should be noticed, that the proposed earlier array modification consisting of introducing the register, which has direct access to each PE, has some crucial disadvantages. The big complication can be creation of a suitable addressing schema and communication registers schema [6].

It should be expected that such solution can higher the level of the array complication and thus its price. It is worth trying to change the attitude to the method of making the whole morphological operation in the systolic array.

It could be done simultaneously with introducing the picture to the array. This way the pulsation devoted only to introducing the picture to the array will be avoided. A certain disadvantage is the fact, that in morphological operations there is a correlation among introduced pixels. Doing morphological operation can not cause the change of introduced pixels because it can lead to incorrect results. Additionally, as the result of undertaken particular morphological operation there is simultaneous existence in registers of pixels from picture before the operation and pixels of picture after the operation. It is possible to make such an operation demanding this data. An example of such operation is shape detection. In case of traditional approach to the processing first the picture to the array should be introduced, further the erosion operation should be done and later its result should be diminished from the first picture to get the contour of the object as the result.

It should be noticed for example that for 4x4 size array and picture 4x4 size four pulsations for introducing the picture to the array will be needed. Next, four or more pulsation, (depending on the class of algorithm) to make an erosion operation and further four or more pulsation to diminish are needed.

Hence, the solution of creating the array that can be used as a filter through which each of the elements of the picture is moved can be created.

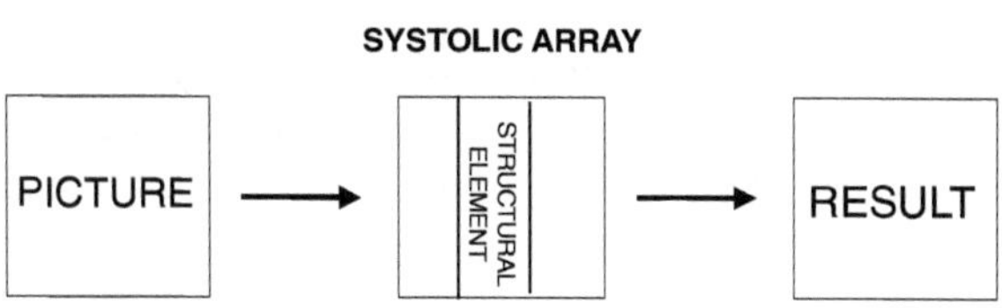

Fig. 10 Systolic array with introduced structural element [10]

Such the solution is optimal because of the time used for the operation. This approach will be described in further dissertations.

7 Designing of Systolic Implementation of Mathematical Morphology Processing

In the process of designing of systolic implementation of mathematical morphology processing transform from "classic" algorithms to their systolic version have to be done. In Fig. 11 diagram of general mathematical morphology algorithm is presented. [16]

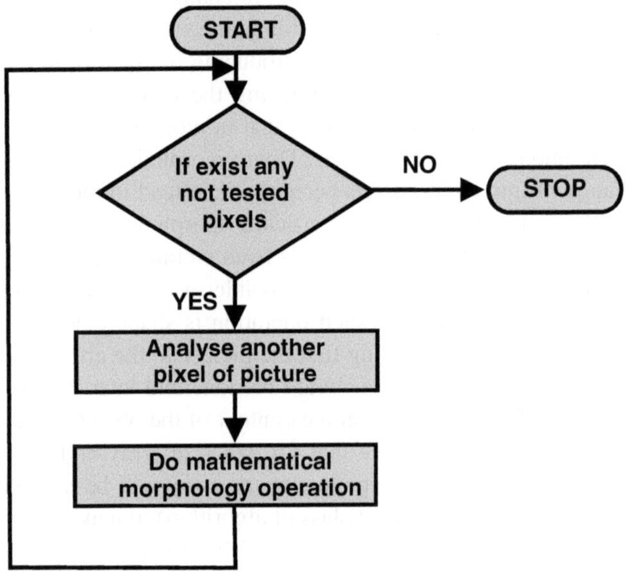

Fig. 11 Diagram of general mathematical morphology algorithm [10]

It should be noticed that transformation of source image is a set of certain operations made on each pixel of image. The systolic array is useful tool to very fast making such operations. Diagram of general mathematical morphology systolic algorithm is presented in figure 12.

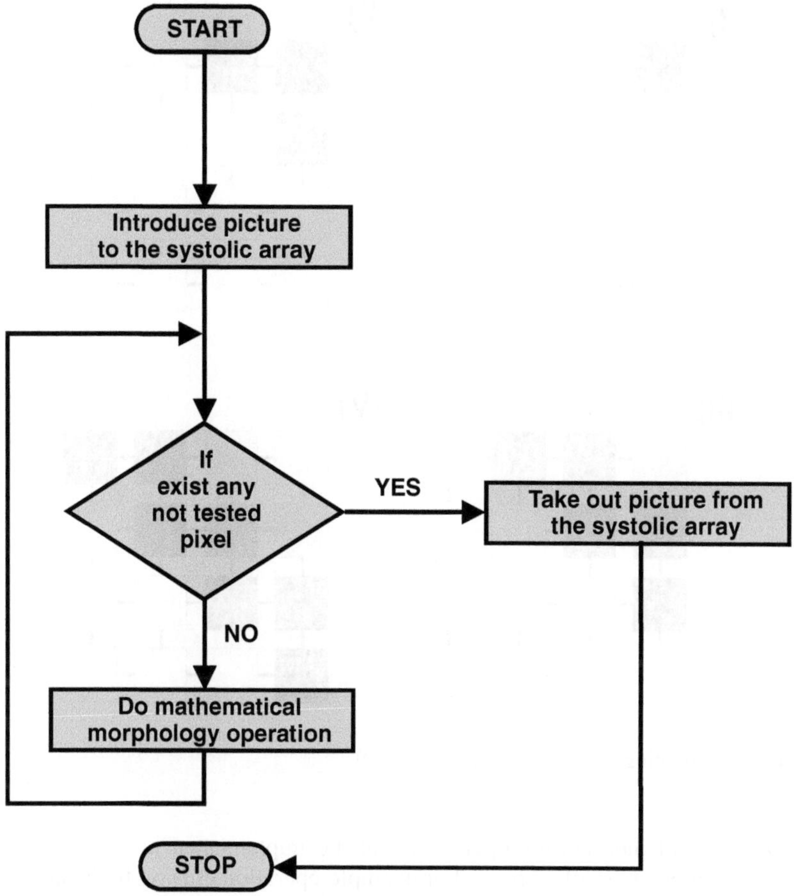

Fig. 12 Diagram of general mathematical morphology systolic algorithm [10]

Individual steps from algorithm presented on figure 12 will be presented below.

It is established that all algorithms will relay on fact that picture is introduced to the array as it is shown on figure 4. The idea of introducing data have to be adjusted to the hardware abilities of systolic array. For example in the Isatec's Systola 1024 array the "north" or "west" direction only could be taken into consideration.

Morphological operations as it was mentioned are characterized by doing set of instructions for every pixel of picture. Idea of floating instructions in systolic array helps to spread morphological instructions in the whole array. Thanks to this following instructions will be done by all PE in array. On figure 18.3 systolic instructions propagation in hypothetical 4x4 systolic array is showed.

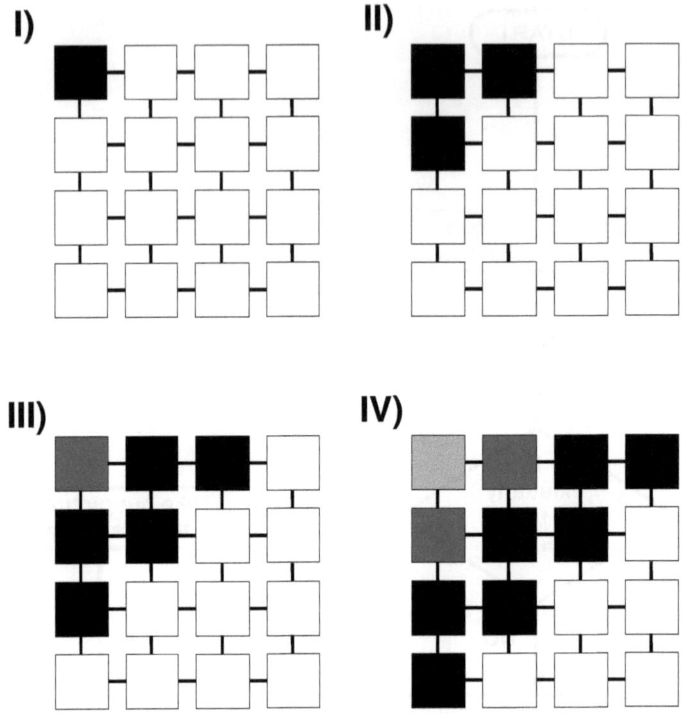

Fig. 13 Data "floating" In systolic array (4x4) [10]

Mathematical morphology operations can be implemented and optimized for specific features of systolic arrays. For example operation erosion finds the lowest value from the picture elements situated under the structural element. Because of the idea of systolic propagations instructions and data in array every processing element "knows" the value of pixels stored in the neighborhood processing elements. It makes easy to processing element to know all values covered by SE needed to make morphological operation.

On figure 13 example pulsation of systolic algorithm is presented. Letters A,..P are the symbols of pixels introduced to the systolic array. Processing elements coordinates 3,1; 2,2 and 1,3 "knows" values stored in the registers of neighborhood processing elements and is able to execute morphological operations. As the result of this operation the pixel of result picture is calculated (on the figure signed as "x").

It should be noticed that every morphological operation consists of operation erosion and/or operation dilatation. For example widely used in pattern recognition Because of the idea of systolic propagation data and instructions the systolic array is an ideal tool for pipelined execution of morphological operations.

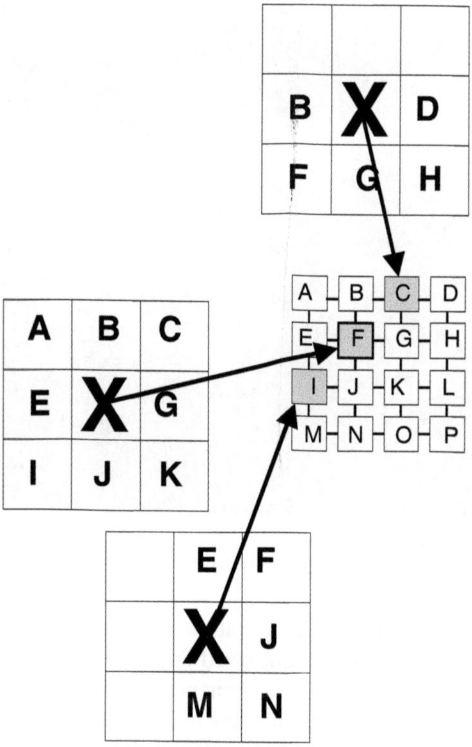

Fig. 14 Example of data stored in registers [10]

In systolic programming it is easy to say whether every pixel of picture is processed. There is no need to make special code to check it. Knowing the size of the systolic array is easy to know how many pulsations are needed to make all the processing operations.

Taking out the result data is making the same way as their introduction. For example in Systola 1024 the side from which data were introduced is the optimal direction. In mentioned authors' proposition of array the optimal direction is opposite to the direction of introduced data.

8 Examples of Usage

Mathematical morphology algorithms can be used to convert pictures to form in which it is easy to find searched feature. Examples of this conversion will be shown below. On figure 15 it is hardly to see rusty places.

Fig. 15 Painted surface before detection and localizations of defects [10]

To find and localize defects the operation erosion has been done twice in the source picture. The structural element sized 3x3 was used. Next operation dilatation has been done (twice). Structural element was the same size as in the erosion operation. Finally the operation erosion has been done. Result of operations is presented in the figure 16.

Fig. 16 Result picture with localized defects (red mark) [10]

To make picture more "smooth" and to find more precisely the defects the operation erosion with structural element sized 3x3 has been done twenty times. Next the operation dilatation has been done twenty times. The result picture is presented on figure 17. The combination of these operations is much more faster in systolic architecture the mathematical morphology open operation with structural element bigger than 3x3.

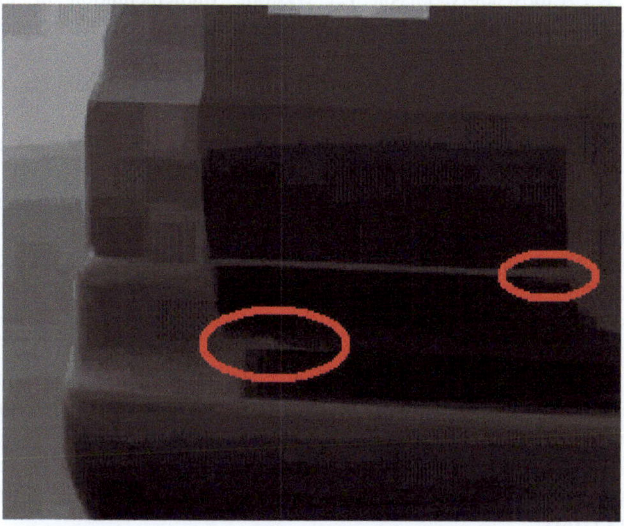

Fig. 17 Result picture with localized defects (red mark) [10]

Presented operations showed that mathematical morphology operations implemented on systolic array are fast and useful tool for searching defects on the pictures [10].

9 Conclusion

To sum it up a well constructed systolic array can be a suitable tool for morphological operations based on the picture. Its additional advantage is certain universality and thus the possibility of making other operations on the picture. It seems that with the present dynamic increase of the amount of digital cameras and video cameras there is some space for the development of systems dedicated to the image processing. It could be filled by systolic arrays.[10],[14],[17]

References

[1] Jain, A.K.: Fundamentals of digital image processing. University of California (1989)

[2] Kung, H.T., Leiserson, C.E.: Systolic Arrays for (VLSI). In: Proc. Sparse Matrix Symp. SIAM (1978)

[3] Kung, S.Y.: VLSI Array Processors. Prentice-Hall Int., Inc. (1988)

[4] Lichy, K., Lipowska-Nadolska, E.: Morphological operations on binary pictures using systolic array Systola 1024. In: Conference ZKwE 2004 (April 2004)

[5] Lichy, K., Lipowska-Nadolska, E.: Systolic implementation of morphological processing – theory and applications. In: XXVII International Conference IC-SPETO 2004, Gliwice-Niedzica (2004)

[6] Lipowska-Nadolska, E., Morawski, M.: Systolic arrays. Akademicka Oficyna Wydawnicza PLJ (1999)

[7] Malina, W., Ablameyko, S., Pawlak, W.: Fundamentals of digital picture processing. Akademicka Oficyna Wydawnicza Exit, Warszawa (2002)

[8] Nieniewski, M.: Mathematical morphology in picture processing. Akademicka Oficyna Wydawnicza PLJ (1998)

[9] The ISATEC Parallel Computer Systola 1024 Documentation, ISATEC Soft- und Hardware GmbH

[10] Lipowska-Nadolska, E., Kwapisz, M., Lichy, K.: Systoliczne przetwarzanie sygnałów cyfrowych. Exit Warszawa (2007)

[11] Polanowski, P., Jung, J., Kiełbik, R., Napieralski, A., Lichy, K.: Od algorytmu dynamicznej cieczy sieciowej do dedykowanego komputera równoległego. Przegląd elektrotechniczny (November 2008)

[12] Bertil, S., Schimmler, M., Heiko, S.: A programmable systolic device for image processing based on mathematical morphology. Computers and Artificial Intelligence 13 (1995)

[13] Qian, K., Lu, X.: An algebraic approach for morphological operations on 2D and 3D image. Pattern Recognition (2008)

[14] Diamantaras, K.I., Kung, S.Y.: A Linear Systolic Array for Real-Time Morphological Image Processing. Journal of VLSI Signal Processing 17(1) (September 1997)

[15] Fejes, S., Vajda, F.: A data-driven alghorithm and systolic architecture for image morphology, `http://umd.edu/~fejes/papers/icip94.ps.gz`

[16] Serra, J.: Image analysis and mathematical morphology, vol. 1. Academic Press, London (1988)

[17] Yen, D.W.I., Kulkarni, A.V.: Systolic processing and an implementation for signal and image processing. IEEE Transations Comput. 31(10) (1982)

[18] Zabrodzki, J.(red.): Grafika komputerowa. Metody i narzędzia, WNT (1994)

[19] Zmuda, M.A., Tambourino, L.A., Rizki, M.M.: Efficient software techniques for morphological image processing with desktop computers. In: Proceedings of SPIE Conference (April 1991)

Adaptive Wavelet Synthesis for Improving Digital Image Watermarking

Jan Stolarek

Abstract. Discrete Wavelet Transform is one of the most popular tools of digital signal processing. Many different wavelet functions have been proposed so far, however there is no wavelet that would be the most suitable for every task. Therefore a method allowing to adaptively synthesize the most suitable wavelet for a given task must be developed. In this paper a general outline of such method will be discussed. A concept of tools used for analysis of adaptive wavelets will be presented.

1 Introduction

The Discrete Wavelet Transform (DWT) became one of the most popular tools in the area of digital signal processing during the last two decades. So far different wavelet families have been proposed, such as Daubechies, Coiflet, Morlet or Mexican hat wavelets [1, 2]. Each of these wavelet functions has its unique properties, making some wavelets more suitable for particular tasks than others. This raises a problem of selecting the best wavelet for a particular task, which leads to a question: does there exist a wavelet that would be the best for the given task, but has not been proposed yet? To provide a general answer to this question a method for automatic adaptive synthesis of wavelets should be developed.

The first step towards wavelet adaptation is parameterization of wavelet filter coefficients. Many such parameterizations have been proposed so far [3,4,5,6]. In papers by Rieder et. al [15] and Vaidyanathan [16] a two channel perfect reconstruction finite impulse response filter was implemented using a parameterized lattice structure. A crucial element in these methods was the optimal selection of parameters' values, which was done using well-known numerical optimization methods. In general, filters (including wavelet filters) have to be adapted for a particular application in mind, and therefore optimality criteria have to be defined. These criteria depend on the problem being solved. That problem is presented below.

During the last 15 years, the Internet has given a rise to the problem of extensive copyright violation. Digital data can be easily copied from one computer

Jan Stolarek
Technical University of Łódź, Institute of Information Technology
Wolczanska 115, 90-924 Lodz, Poland
e-mail: stolarek@ics.p.lodz.pl

P. Lipiński, K. Świrski (Eds.): Towards Modern Collaborative Knowl. Sharing Syst., SCI 401, pp. 133–143.
springerlink.com

to another. It is very hard to prove one's ownership of a particular digital work, e.g. image. As a solution to that problem digital data watermarking [7] was created. In this paper digital image watermarking will be discussed. The concept of that method is to embed additional data into the digital image. This additional data – a watermark – is used to prove the ownership of the image. Therefore a watermark should be impossible to remove without damaging the image beyond usability. Moreover, embedding the watermark shouldn't alter the image significantly, as not to interfere with the normal image usage. In the recent years digital watermarking in the wavelet transform domain has gained much popularity. This is caused by the good time–frequency localization properties of the Discrete Wavelet Transform (DWT), which allows to embed a watermark only in the selected regions and frequencies of an image. So far, the authors of the watermarking algorithms have been arbitrarily choosing the wavelet used for image decomposition. Some works have been presented, that introduce a concept of wavelet filter parameterization to improve the digital watermark security – e.g. in paper by Dietl et al. [14] random selection of parameters was proposed to increase the watermark robustness against attacks.

In this paper we discuss a concept of parameterizing wavelet filters using orthogonal lattice structure and performing adaptive parameter selection to adapt the wavelet and thus increase the watermark embedding robustness (sepearability and resistance to attacks) and watermarked image fidelity (quality of the watermarked image). In the previous research it was already demonstrated, that such adaptation is possible and can be carried out using the genetic algorithm approach [8, 9].

This paper is organized as follows. Wavelet filter parameterization using the orthogonal lattice structure is presented in Section 2. The theory behind the digital image watermarking in the wavelet transform domain is presented in Section 3. Section 4 discuses the details of genetic approach to optimization of the lattice structure parameters and presents the optimality criteria that will be used for the wavelet synthesis. Section 5 presents a concept of tools for studying the relation between the wavelet filter parameters and the effectiveness of synthesized wavelets. A summary and the direction of future research are given in Section 6.

2 Orthogonal Wavelet Transform

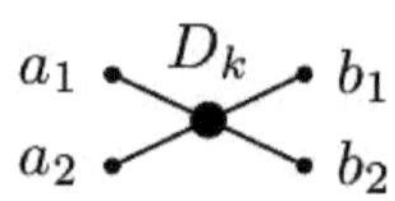

Fig. 1 Base operation

Orthogonal lattice structure is a computational scheme introduced in [10] and discussed in more detail in [11]. It is based on two point base operations (presented in Figure 1), that can be viewed as a 2x2 matrix multiplication:

$$\begin{bmatrix} b_1 \\ b_2 \end{bmatrix} = D_k \cdot \begin{bmatrix} a_1 \\ a_2 \end{bmatrix}, \text{ where } D_k = \begin{bmatrix} w_1^{(k)} & w_2^{(k)} \\ w_2^{(k)} & -w_1^{(k)} \end{bmatrix}, \tag{1}$$

It is assumed, that D_k is orthogonal:

$$D_k \cdot D_k^T = I, \tag{2}$$

which means that weights $w_1^{(k)}$ and $w_2^{(k)}$ fulfil the following conditions:

$$w_1^{(k)} w_2^{(k)} - w_1^{(k)} w_2^{(k)} = 0, \tag{3}$$

$$(w_1^{(k)})^2 + (w_2^{(k)})^2 = 1. \tag{4}$$

Condition (3) is of course always satisfied, which is ensured by the definition of the D_k base operation. Equation (4) is explicitly satisfied, when the following substitutions are assumed [12]:

$$\mathrm{w}_1^{(\mathrm{k})} = \cos(\alpha_k), \ \mathrm{w}_2^{(\mathrm{k})} = \sin(\alpha_k). \tag{5}$$

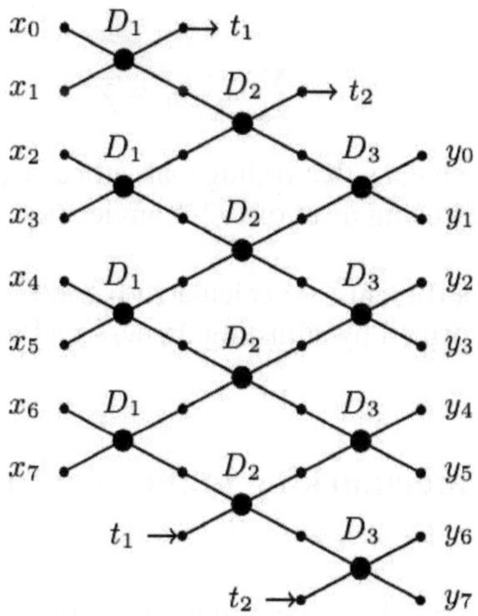

Fig. 2 Lattice structure implementing 6-tap transform

Base operations are composed into layers (see Figure 2). Orthogonal lattice structure is composed of *K/2* stages, each containing D_k operations repeated *N/2* times, where *K* and *N* are the lengths of the wavelet filter's impulse response and of a processed signal respectively. On each stage of the lattice structure, elements of the signal are processed in pairs by D_k base operations. After each stage, base operations are shifted down by one and a lower input of the last base operation in the current stage is connected to the upper output of the first base operation in the preceding stage (t_1 and t_2 in Fig. 2). Upper outputs of base operations in the last layer (y_0, y_2, y_4 and y_6 in Fig. 2) correspond to the low–pass filter signal and lower outputs (y_1, y_3, y_5 and y_7 n Fig. 2) correspond to the high–pass filter signal.

A lattice structure, with base operations fulfilling conditions (2)–(4), is able to implement a wavelet filter bank that fulfils conditions of wavelet decomposition, assuming that the following constraint imposed on the α_k angles holds true [17]:

$$\underset{n\in Z}{\exists} \sum_{l=1}^{L/2} \alpha_l = \frac{\pi}{4} + \left[\left(\frac{L}{2} - 1 \right) \bmod 4 \right] \cdot \frac{\pi}{2} + 2n\pi . \tag{6}$$

To ensure that the above equation always holds true, the following representation has been introduced [15]:

$$\begin{gathered} \alpha_1 = \gamma - \varphi_1 \\ \alpha_i = (-1)^i (\varphi_{i-1} - \varphi_i), \text{ for } i = 2, \ldots, \frac{L}{2}, \\ \alpha_{L/2} = (-1)^{L/2} \varphi_{L/2-1} \end{gathered} \tag{7}$$

where γ is the desired sum of angles and is determined depending on the number of layers in the orthogonal lattice structure. Such a representation assures, that condition

$$\sum_{l=1}^{L/2} \alpha_l = \gamma , \tag{8}$$

is always fulfilled. It also means that orthogonal lattice structure, consisting of L/2 layers, can be represented using a set of L/2-1 angles ($\varphi_1,\ldots,\varphi_{L/2-1}$), instead a set of L/2 angles ($\alpha_1,\ldots, \alpha_{L/2}$).

Wavelet filter bank coefficients are calculated based on the D_k base operations. Wavelet synthesis is performed by adjusting the $w_1^{(k)}$ and $w_2^{(k)}$ parameters of the D_k base operations.

3 Digital Image Watermarking in the Wavelet Transform Domain

Many digital image watermarking algorithms operating in the wavelet transform domain have been proposed so far. All of these algorithms share a common embedding scheme. The original data is first decomposed using the discrete wavelet transform and the Mallat algorithm for multilevel decomposition of a signal (usually 3 to 5 image decomposition levels are used). The watermark is then embedded in the wavelet coefficients using the selected embedding algorithm. Inverse wavelet transform is applied to reconstruct the watermarked image. To extract the watermark, image decomposition using discrete wavelet transform must be performed on the watermarked image. Watermark extraction algorithm is then used to recover the watermark from the wavelet coefficients. The extracted watermark is then compared to the embedded watermark. If their similarity (in most cases measured as correlation) is large enough, the watermark presence is detected.

The approach proposed in this paper operates on the generic watermark embedding method called E_BLIND [7], but it can also be applied to any other method of digital image watermarking in the wavelet transform domain. In the E_BLIND algorithm the watermark w_r is a random sequence of N integer numbers of the set $\{-1, 1\}$. Multilevel Discrete Wavelet Transform of the image is performed using Mallat's decomposition algorithm. N largest wavelet coefficients from all three detail sub–bands on third level of image decomposition are selected. This means, that the watermark is embedded in the mid–frequencies of an image. The watermark is embedded in the selected coefficients using formula

$$c_w = c_0 + \alpha |c_0| w_r , \tag{9}$$

where c_0 are the selected wavelet coefficients, α is the embedding strength, w_r is the watermark and c_w are the watermarked wavelet coefficients.

To extract the watermark, watermarked image has to be decomposed using the Mallat's algorithm with the same number of decomposition levels as was used for watermark embedding. Location of the watermarked coefficients must be known. Watermark is detected by computing the normalized correlation between the watermarked wavelet coefficients and the original watermark according to the formula

$$C = \frac{1}{N-1} \sum_{k=1}^{N} \frac{(c_w(k) - \overline{c_w})(w_r(k) - \overline{w_r})}{\sigma_r \sigma_w} , \tag{10}$$

where N is the length of the watermark, c_w denotes the watermarked coefficients, $\overline{c_w}$ is the mean value of the watermarked coefficients, w_r denotes the embedded watermark, $\overline{w_r}$ is the mean value of the embedded watermark, σ_c and σ_w are standard deviations of the watermarked coefficients and the watermark respectively. Presence or absence of the watermark is usually determined with a threshold τ. If the correlation C is greater than τ then the watermark is present, otherwise it is absent. It is therefore important to maximize the correlation.

4 Wavelet Adaptation Using the Genetic Algorithm

When the base operations of a lattice structure are modified, the output signal from the lattice structure changes. This signal can be rated in terms of its fitness in respect to some quality criteria. In the digital image watermarking, there are three contradicting fitness criteria:

- Correlation between the extracted watermark and the random watermarks should be minimized, while maximizing the correlation between the extracted watermark and the embedded watermark (separability),
- Visual difference between the original image and the watermarked image should be minimized (fidelity),
- Watermark resistance against attacks should be maximized (robustness).

This turns the problem of mother wavelet synthesis using lattice structure into a multi-objective optimization problem, which can be solved using the genetic algorithm approach. The author proposes an algorithm based on Simple Genetic Algorithm and expanding it to use Evolution Strategies. Below is the outline of proposed genetic algorithm:

- initialize random population P of μ individuals
- FOR k = 1 TO iterations_count DO
- evaluate fitness of individuals in population P
- create temporary population T containing λ individuals selected from population P
- perform crossover and mutation on individuals in population T
- evaluate fitness of individuals in population T
- select μ individuals to form new population P
- END FOR
- display the best individual in population P

The stop condition is reaching the maximum number of iterations. Individuals in the genetic algorithm are represented as a set of binary coded $(\varphi_1,\ldots,\varphi_{L/2-1})$ angles, representing the orthogonal base operations (according to equations (1), (4) and (6)). Each φ_k angle is binary coded on a chromosome ch_k using m bits. This means that it can be treated as an integer from range $[0,\ 2^m)$. To convert chromosome ch_k to φ_k angle the following formula is applied:

$$\varphi_k = 2\pi\frac{ch_k}{2^m} = \frac{ch_k \cdot \pi}{2^{m-1}}. \tag{11}$$

It assures that φ_k angle falls into range $[0,2\pi)$. After all the φ_k angles have been calculated, they are converted to α_k angles using equation (6). These angles are used to calculate the base operation according to equations (4) and (1). After the base operation are calculated they can be used to perform the wavelet transform in the watermark embedding algorithm.

To evaluate individual's fitness in terms of criterion 1 a set of random watermarks must be generated. First, the normalized correlation between the extracted watermark and the embedded watermark is calculated. Then the normalized correlation between the extracted watermark and the random watermarks is calculated. The fitness of i–th individual in terms of the first criterion is calculated using the formula:

$$F_i^{(1)} = \min_j (C_e^{(i)} - C_{rj}^{(i)}), \tag{12}$$

Where $C_e^{(i)}$ is the normalized correlation between the extracted watermark and the watermark embedded using the i-th individual (filter) and $C_{rj}^{(i)}$ is the normalized correlation between the extracted watermark and the j-th random watermark for i-th individual.

Fitness in terms of the second optimality criterion is calculated using the standard image similarity assessment method – PSNR:

$$F_i^{(2)} = 10 \cdot \log_{10} \frac{255^2}{MSE}, \tag{13}$$

where MSE is the Mean Square Error between the watermarked image and the original image:

$$MSE = \frac{1}{NM} \sum_{i=1}^{M} \sum_{j=1}^{N} [O(i,j) - W(i,j)]^2, \tag{14}$$

where O(i,j) and W(i,j) are pixel values of original and watermarked image respectively, N and M are image dimensions.

The optimality in terms of the third criterion isn't assured by directly computing individual's fitness. Instead, after the watermark is embedded in the image a certain type of image attack (scaling, compression, low-pass or median filtering, etc.) is performed on the watermarked image before the watermark is extracted. Then the fitness $F_i^{(1)}$ is calculated. The performed attack influences that fitness, thus individuals, that have high fitness, despite the attack performed on the image, are discovered.

As the method for multi-objective optimization the Global Optimality Level is used [13]. The first step is determination of the maximal and minimal value in the population of each partial fitness:

$$\begin{aligned} F_{\min}^{(j)} &= \min_i \{F_i^{(j)}\} \\ F_{\max}^{(j)} &= \max_i \{F_i^{(j)}\}, \end{aligned} \tag{15}$$

where i and j are the indices of individuals and partial fitness respectively. The final fitness F_i of an individual is calculated using the formula:

$$F_i = \min_j \left\{ \frac{F_i^{(j)} - F_{\min}^{(j)}}{F_{\max}^{(j)} - F_{\min}^{(j)}} \right\}. \tag{16}$$

This equation means, that every partial fitness of an individual is normalized to fit into range [0,1]. The worst partial fitness is then selected as the final fitness of an individual. Such an approach promotes individuals that ensure both the good image quality and the resistance to the attacks.

To create the temporary population T (step 4 in the algorithm outline), λ individuals are selected from the base P population containing μ individuals, where $\lambda > \mu$. Either the Tournament Selection or the Roulette Wheel Selection can be applied. Individuals in the temporary population are modified using two genetic operators: the crossover operator and the mutation operator. These operators create new individuals (solutions) within the population *T*. Having the original population *P* and the temporary population *T*, the new population *P* must be created. Author proposes to use the evolution strategies for that purpose: $(\mu + \lambda)$ strategy and (μ, λ) strategy. In the $(\mu + \lambda)$ strategy, the new population *P* is created by selecting μ fittest individuals from both the parents (population *P*) and the offspring (population *T*). This strategy assures stability of the optimization process – if the new individuals from population *T* are less fit than in the original population, they will not be placed into the next generation (the parents are copied to the next generation if they have higher fitness). This, however, may lead to dominating the population by fit individuals early in the optimization process, and thus lead to premature convergence. In the (μ, λ) strategy the original population *P* is discarded and the new population *P* is created by selecting μ fittest individuals from the *T* population. Experiments have shown, that if no evolution strategy is applied, then unfit individuals are not eliminated from the population, which leads to destabilization of the optimization process.

Genetic algorithms are an efficient method of global optimization for solving complex tasks and have proven to be an effective method of wavelet synthesis for improving digital image watermarking [9].

5 Tools for Adaptive Wavelet Synthesis

So far there have been no tools that would allow to adaptively synthesize wavelets nor there have been tools to study the influence of wavelet parameters on the watermarking process. Therefore the author of this paper proposes to create two separate tools to aid the research of adaptive wavelets and their influence on the digital image watermarking process. These tools are discussed below.

5.1 Wavelet Plotter

Wavelet Plotter will be the basic tool to study the influence of orthogonal lattice structure parameters on the digital image watermarking process. Its basic functionality is plotting of the scaling function based on the parameters given by the user (the parameters for Daubechies 4, Daubechies 6 and Daubechies 8 wavelets are stored in the program). Coefficients of the filter will be displayed in the program and the plot precision can be adjusted. The second of the Wavelet Plotter's functionalities is embedding the digital image watermark. User will be able to select the image file as well as the embedding parameters (watermark amplitude, watermark length etc.). Watermark embedding method will be E_BLIND, although it will be easy to extend the program with other watermarking algorithms. The watermarked image will be displayed, so the user will be able to visually rate the watermarking fidelity. Peak Signal-to-Noise Ratio (PSNR) of the watermarked image will also be displayed to the user. Beta version of the application in Polish language is presented in the Figure 3.

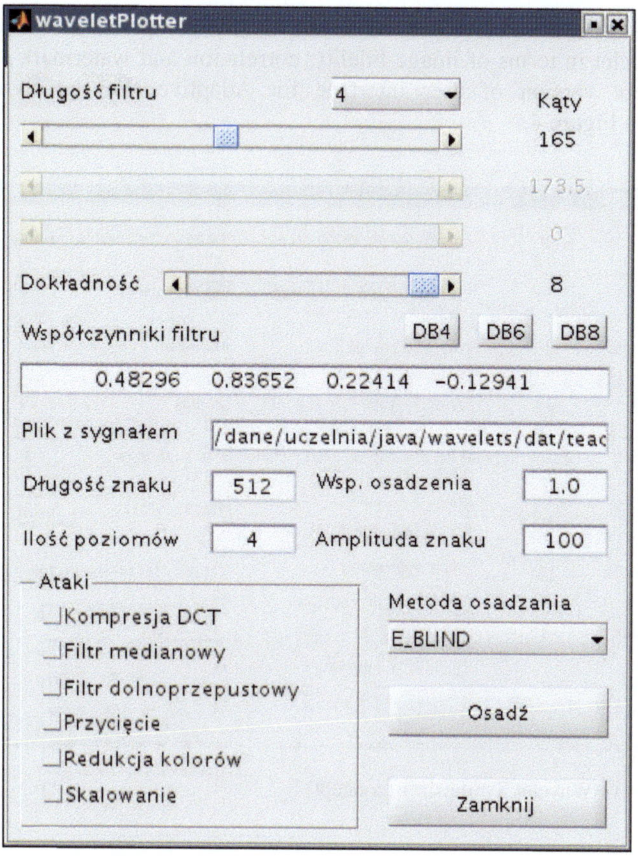

Fig. 3 Wavelet Plotter (beta version)

5.2 Adaptive Wavelet Synthesizer

While Wavelet Plotter allows to manually adjust the wavelet parameters, the second of the proposed tools will perform automatic adaptive wavelet synthesis using a genetic algorithm. User will be able to adjust the parameters related with the wavelet decomposition (filter length, number of wavelet analysis levels), the genetic algorithm (number of algorithm's iterations, representation precision, crossover and mutation probability, selection method, crossover method, evolutionary strategy, population size) and the watermarking process (watermark length and amplitude, embedding strength, embedding algorithm). It will be also possible to perform various kinds of attacks on the watermarked image (DCT compression, median and low–pass filtering, scaling, colour depth reduction) to synthesize wavelets that will make the watermark resistant to these attacks. As a final result of the program, user will be provided with the low–pass and high–pass filter coefficients of the best synthesized wavelet. Moreover, the effectiveness of the synthesized wavelet will be automatically compared to the Daubechies family wavelet in terms of image fidelity, correlation and watermark separability. The concept version of user interface for Adaptive Wavelet Synthesizer is presented in Figure 4.

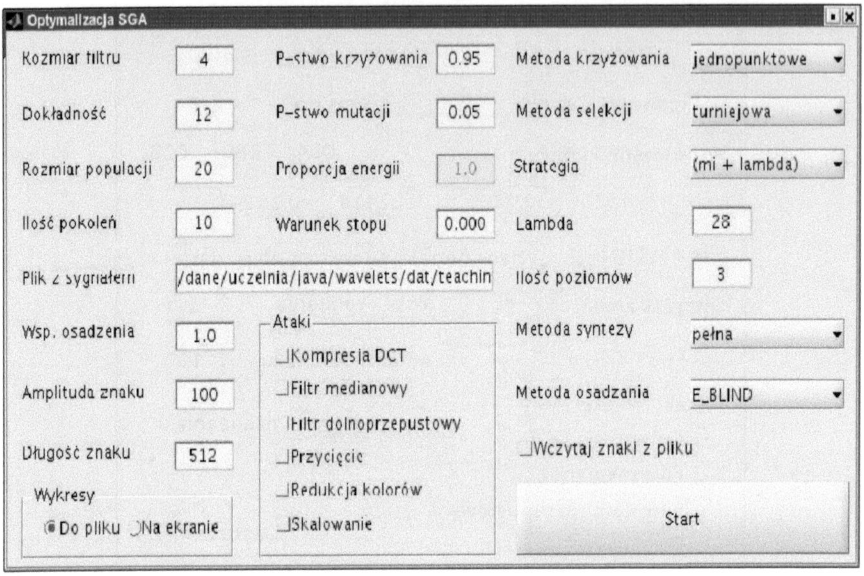

Fig. 4 Adaptive Wavelet Synthesizer (concept)

6 Summary

In this paper an orthogonal lattice structure used for wavelet parameterization was presented. The problem of improving digital image watermarking effectiveness was discussed. During the earlier research it was shown, that adjusting the wavelet function can significantly improve the watermarked image fidelity and the watermark separability. Two applications have been proposed to aid the scientists researching the adaptive wavelet synthesis. First of the applications will allow to manually adjust wavelet parameters to study the particular wavelet function. Second of the applications will automatically synthesize adaptive wavelets to study how the watermark length, the embedding strength and the attacks performed on the image influence the coefficients of synthesized wavelets.

As part of the future research both applications will be implemented in the Matlab environment. Created programs will be used to further study the practical aspects of digital image watermarking improvement. Presented application concepts may be further expanded, e.g. different image quality assessment methods or meta algorithms for auto selecting parameters of the evolutionary algorithm may be implemented. Moreover, the theoretical background of proposed lattice structure will also be studied in more detail (first part of research has already been published in [17]).

References

[1] Daubechies: Ten Lectures on Wavelets. SIAM (1992)

[2] Mallat, S.: A wavelet tour of signal processing. Academic Press (December 2008)

[3] Lipiński, P., Yatsymirskyy, M.: On synthesis of 4-tap and 6-tap reversible wavelet filters. Przegląd Elektrotechniczny (12), 284–286 (2008)

[4] Regensburger, G.: Parametrizing compactly supported orthonormal wavelets by discrete moments. Applicable Algebra in Engineering, Communication and Computing 18(6), 583–601 (2007)

[5] Odegard, J.E., Burrus, C.S.: New class of wavelets for signal approxima-tion. In: IEEE International Symposium on Circuits and Systems (ISCAS) (May 1996)

[6] Wei, D., Bovik, A.C., Evans, B.L.: Generalized coiflets: a new family of orthonormal wavelets. In: Record of the Thirty-First Asilomar Conference on Signals, Systems & Computers, November 1997, vol. 2, pp. 1259–1263 (1997)

[7] Cox, I.J., Miller, M.L., Bloom, J.A., Fridrich, J., Kalker, T.: Digital Water-marking and Steganography, 1st edn. Elsevier (2008)

[8] Stolarek, J., Lipiński, P.: Improving digital watermarking fidelity using fast neural network for adaptive wavelet synthesis. Journal of Applied Computer Science 18(1), 61–74 (2010)

[9] Stolarek, J., Lipiński, P.: Digital watermarking enhancement using wavelet filter parametrization (submitted for publication)

[10] Yatsymirskyy, M.: Lattice structures for synthesis and implementation of wavelet transforms. Journal of Applied Computer Science 17(1), 133–141 (2009)

[11] Stolarek, J.: Adaptive synthesis of a wavelet transform using neural network. Bulletin of Polish Academy of Sciences, Technical Sciences 59(1) (March 2011)

[12] Stasiak, B., Yatsymirskyy, M.: Fast Orthogonal Neural Networks. In: Rutkowski, L., Tadeusiewicz, R., Zadeh, L.A., Żurada, J.M. (eds.) ICAISC 2006. LNCS (LNAI), vol. 4029, pp. 142–149. Springer, Heidelberg (2006)

[13] Kowalczuk, Z., Białaszewski, T.: Genetic algorithms in multi-objective optimisation of detection observers. In: Korbicz, J., Kościelny, J.M., Ko-walczuk, Z., Cholewa, W. (eds.) pp. 511–556. Springer (2004)

[14] Dietl, W., Meerwald, P., Uhl, A.: Protection of wavelet-based watermark-ing systems using filter parametrization. Signal Processing 83(10), 2095–2116 (2003)

[15] Rieder, P., Gotze, J., Nossek, J.S., Burrus, C.S.: Parameterization of or-thogonal wavelet transforms and their implementation. IEEE Transactions on Circuits and Systems II: Analog and Digital Signal Processing 45(2), 217–226 (1998)

[16] Vaidyanathan, P.P., Hoang, P.-Q.: Lattice structures for optimal design and robust implementation of two-channel perfect-reconstruction QMF banks. IEEE Transactions on Acoustics, Speech and Signal Processing 36(1), 81–94 (1988)

[17] Stolarek, J.: On properties of a lattice structure for a wavelet filter bank implementation: Part I. Journal of Applied Computer Science 19(1) (2011)

Congestion Avoiding Flow Control Solution Applied to Bounded Area of IP Network

Michał Karbowańczyk

Abstract. In this chapter a simple, credit-based flow control solution for IP networks is presented. In contrast to several proposals that utilise flow-level control (where control rules are applied to the source of considered flow), our solution is designed as link-level control system, where every set of directly connected network nodes creates a separate control system. This approach allows the network administrator to apply proposed solution to selected area of IP network. It is particularly important in case of the network that the end (leaf) systems are out of control of network administrator (i.e. mobile systems), thus the flow control mechanism cannot be applied to them. The results of experimental research in real IP network show that in such case the proposed solution is able to avoid congestion and noticeably improve the effectiveness of data transmission within considered network area.

1 Introduction

As the spectrum of applications where Internet is involved, in business as well as entertainment domains, still grows, we observe that the demand for higher transmission rates and reliability is also rising. To meet these requirements, different flow control solutions, that allow the data sources to adopt their data sending rate to actual capabilities of network links and intermediate nodes (in case of Internet called routers), are applied. One of the most important goals of traffic control is to avoid congestion of network nodes. By the term "congestion" we mean situation, when memory buffer that considered node utilises to store the queue of data units (i.e. IP packets) which are to be transmitted, fills up. In such case the node is not able to hold subsequently received data units, so it has to drop them. If data integrity is guaranteed by the flow that the dropped data units belonged to, the source of that flow is obliged to emit them again, and that activity is usually called "retransmission". As retransmitted data units traverse several links more than once, it

Michał Karbowańczyk
Technical University of Łódź, Institute of Information Technology
Wolczanska 215, 90-924 Lodz, Poland
e-mail: mak@zsk.p.lodz.pl

P. Lipiński, K. Świrski (Eds.): Towards Modern Collaborative Knowl. Sharing Syst., SCI 401, pp. 145–156.
springerlink.com

can be obviously said that utilization of the available throughput of these links is less effective. Moreover, retransmissions lead to increase amount of data units that arrive at the congested node, so the problem of the congestion escalates.

The overall goal of our research work is to provide a complete traffic control solution that could be used Internet-wide or within a selected network segment, with no need to modify the design of existing protocols nor the way that existing applications make use of them. In [11] we provided extensive arguments that substantiate the need of developing new solutions of that problem; below we present summarised conclusions of mentioned discussion.

The de facto standard of Internet congestion avoidance and control is the congestion window (CWND) mechanism utilised by Transmission Control Protocol (TCP). There are plenty of approaches that are considered to determine congestion window size, including classical AIMD (Additional Increase, Multiplicative Decrease), heuristic, evolution strategies, optimization methods and many others. An interesting survey on these works is presented in [5]. The common issue related to various TCP congestion control algorithms is that they utilise implicit feedback received from the network, and the term "implicit" generally means that the sender (which runs the algorithm) discovers that the congestion already occurred by noticing that the receiver did not acknowledge the receipt of at least one of the segments that were sent to it. Thus congestion control concept of TCP cannot be considered as the solution of congestion avoidance problem. These remarks are also valid for the Active Queue Management (AQM, shortly named "active queues") proposals, which were introduced in order to improve robustness of TCP congestion window mechanism. Active queue concept particularly aims at addressing one of the most important issues observed when multiple TCP connections are served by a congested node. As the sources simultaneously decrease and then increase their congestion windows, subsequent congestions occur in a periodical manner. Active queue mechanism is able to affect this process by properly selecting the packets that it decides to drop before the buffer becomes overflown. Several AQM algorithms are thoroughly discussed in [3]. Said that, we clearly notice that active queues do not eliminate congestions, although they are utilised to minimise their negative consequences. Furthermore, we have to state that the only reasonable way to completely avoid network congestions is to provide the flow control algorithm with explicit feedback on the actual network state. Unfortunately, this is not addressed in Internet Protocol (IP) nor TCP specification, except rough and not necessarily supported Explicit Congestion Notification (ECN) option. There is some effort made to design a transport layer level traffic control solution that would introduce a closed loop providing explicit feedback, i.e. eXplicit Control Protocol (XCP) [10]. Despite of desirable features of such protocols, it has to be claimed that their deployment would be possible only if existing applications were redesigned to support them, what in turn is inconsistent with the aim of our work.

The rest of this chapter is organised as follows. In section 2 we briefly describe the important assumptions and design considerations of the proposed solution. For those interested in thorough discussion on the concept and implementation of closed loop providing explicit feedback or the details and mathematical analysis of control algorithm, we suggest reading [11] and [4], respectively. In section 3 we present topology, setup and results of experiment carried out on a real IP network. This experiment was focused on examining the effectiveness of the proposed flow control solution applied to selected area of considered network. The results that we obtained are then compared with similar tests performed in case when the flow control is applied to every node of considered network, or it is not utilised at all.

2 Flow Control Solution

2.1 Control Algorithm

The proposed flow control solution utilises discrete time control algorithm presented in details in [4]. It is run by the controller at time instants $k \cdot T$, where T denotes the discretization period and k numbers subsequent cycles. The determined control value designates amount of data that the controlled object is allowed to send. It is calculated with following formula

$$a(k) = X^D - x(k) - w(k)$$

where: $a(k)$ denotes control value, $x(k)$ denotes the length of the queue stored by node in its buffer at time instant $k \cdot T$, and $w(k)$ denotes the sum of control values that were already sent to the controlled nodes, but they have not affected the state of the controller node yet (the controlled nodes did not so far receive them, or they stored them for future use), at the same time instant. X^D is a reference value chosen by network administrator. This value also specifies the minimum capacity of the controller node buffer, as the length of the packet queue never exceeds it, what is proven in [4]. Thus X^D obviously has to be positive; in fact, there is no use for this value to be lower than maximum packet size. Furthermore, there is a condition on the reference value formulated in [4]; if it is met, effective utilization of throughput available for considered node is guaranteed.

Taking into account ISO/OSI layered network model, it is very important to properly choose the layer that the control algorithm is to be applied to. The session, presentation and application layers are out of scope of our interest, as they are intended for designing application-level protocols. On the other hand, flow control is usually implemented at transport layer, as this is the level where elementary flow is defined. However, there are two general issues concerning the idea of implementing a new transport layer level solution. First, if we decide to design a new transport layer protocol, the existing applications will have to be modified to support it, as we already mentioned in section 1. Moreover, we have to consider that there are transport level protocols that do not provide any form of flow control (i.e. UDP). In case that adaptive flows (that do perform flow control) share link throughput with non-adaptive flows (that do not), the former tend to excessively reduce their rates, and most of available throughput is consumed by the latter. This issue obviously can be addressed only at the lower layers. One might consider data link layer as the place to implement a control flow algorithm. Indeed, in this approach the control law would be definitively applied to whole traffic transmitted, regardless of higher layers protocols used. Furthermore the control mechanism could utilise specific properties of reflected data link layer technology, such as closed loop providing feedback facility in ATM/ABR virtual connection. Unfortunately, as effective scope of a mechanism designed for a selected data link layer protocol is for natural reasons limited to the homogenous physical network segment, it is not consistent with the assumption that it is possible for the proposed solution to be used Internet-wide. Now take into account the network layer, namely – Internet Protocol. It is evidently deployed worldwide and it is responsible for delivering packets carrying higher layers data, so if we apply a flow control here it will have impact on every elementary flow, regardless of the adaptive or non-adaptive nature of its transport layer protocol. Considering above remarks, we finally decided to design and implement a flow control facility that operates at IP level. To achieve the goal we developed a queuing discipline, which is an extension to the network subsystem of Linux kernel.

2.2 Feedback Loop

An important part of the proposed solution is a feedback loop. It is responsible for delivering control values, calculated by the control algorithm (running on the controller node), to the nodes that are controlled ones. Taking into account a set of nodes that make up a considered flow, there are two general approaches to the problem of establishing a feedback loop.

In a well known flow-level control approach the loop is established between two leaf nodes of the flow, namely source and destination. The source prepares a dedicated data unit called control unit (i.e. Resource Management cell of ATM/ABR network) and sends it to the destination, which in turn sends it back. Any of the intermediate nodes is allowed to put an information on its state into control unit, and the source utilises this information to obtain the desired rate at which it sends data. A commonly used in research works [1,2,4, 6-9] variant of this approach is a bottleneck scenario. In that case one of the intermediate nodes is selected and pointed at as a bottleneck node, and it is responsible for calculating the control value that determines the source behaviour. The term "bottleneck" refers to the assumption that the selected node is characterised by the worst transmission capabilities over all considered configuration. Apart from distinguishing the bottleneck node, this approach causes no changes to the circulation of the control units. Both variants of flow-level control are appropriate for connection-oriented networks, where data transmission must be preceded by a procedure of connection setup that (among others) determines a path – a set of intermediate nodes that take their parts in data transmission. The specific of Internet Protocol is totally opposite, as it belongs to the datagram-oriented class of networks. In such a network, even if we consider a flow controlled by a connection-oriented transport layer protocol (such as TCP), the route from the source to the destination may be different for every single data unit (packet). This leads to the conclusion that even if it is possible to distinguish a bottleneck node and that node is able to determine proper (in terms of the goal of flow control, i.e. congestion avoidance) control value, this value can become inadequate due to route changes that occurred while the control unit was being delivered back to the source.

In the less common link-level control approach the control system consists of two neighbour (connected directly by the link – a mean of lower network layer) nodes. Thus the route of the data unit (or the path of the connection) is decomposed to the set of independent controller node – controlled node pairs. Route changes do not affect the way that particular configurations work. Establishing the control loop between neighbour nodes is also for the benefit of reducing the control delay, which in case of network flow control is concerned with the time that it takes to deliver the control unit to the destination and back to the source. This time delay is usually called *RTT* (Round Trip Time). The common finding is that the lower *RTT* is, the better results a control mechanism can achieve; it is particularly shown in [9-12], where several control algorithms based on Smith's predictor are proposed. These considerations lead us to conclusion that only link-level control approach is appropriate for the problem we are going to solve. Said that, it must be claimed that the specification of Internet Protocol does not provide any facilities that could be utilised to establish closed loops between neighbour nodes of IP network. This problem and its solution is described in detail in [11][12], so we put just a short overview here. A node that acts as a controller must be able to determine a list of elementary flows that traverse it. In case of Linux kernel it is provided by connection tracking module (conntrack). Moreover, for every such a flow, a "previous" (from the perspective of the direction of the flow) neighbour node must be identified. As this information can not be obtained from IP packet header, we decided to distinguish neighbour nodes by their data link layer addresses. To achieve this, we extended the connection tracking module and also created a new packet filter that reads and stores source data link layer addresses of incoming frames.

The communication between controller node and its controlled neighbours proceeds as follows. Just after the control algorithm calculates the control value (using the formula mentioned in Subsection 2.1), a list of elementary flows is gathered from the connection tracking module. As each elementary flow description contains data link layer address of the neighbour node that it passes through, it is possible to enumerate neighbour nodes. The control value is then distributed among the neighbour nodes proportionally to the number of elementary flows assigned to each one. Next, for each neighbour node a control packet is generated. It is an IP packet that contains a control messages sequence defined as follows:

- number of messages (1 byte)
- message 1
    - message type (1 byte). At the moment there is only one message type supported: control value assign (1)
    - message value (4 bytes). If message type is 1, the message value is interpreted as an amount of data that the neighbour node is allowed to send.
- message 2 and so on.

The control packet is encapsulated in a data link layer frame, and the destination address of this frame is set to the determined address of the neighbour node. Finally, the frame is passed directly to the data link layer API so that it is sent to its destination. As a consequence we assume that the successful delivery of the control packet is provided by network infrastructure.

3 Experimental Verification

We examined the proposed solution through a couple of tests that were run in real IP network. The goal of the experiment was to determine, how the solution, deployed at Internet Protocol level, affects the way that several TCP connections share variable, unpredictable link throughput. Every node of the considered network was running Fedora operating system based on vanilla kernel version 2.6.26. By the term "vanilla" we mean that it was downloaded as source code form the official Linux project site [14], extended according to the description provided in section 2, and finally compiled. Two tools available for Fedora: *nc* [15] and *hping* [16] were used as network traffic generators. We decided to utilise the TCP variant known as CUBIC TCP [13], as it is chosen by default since kernel version 2.6.18.

Network Configuration
To perform the tests we set up a topology presented in Fig. 1.

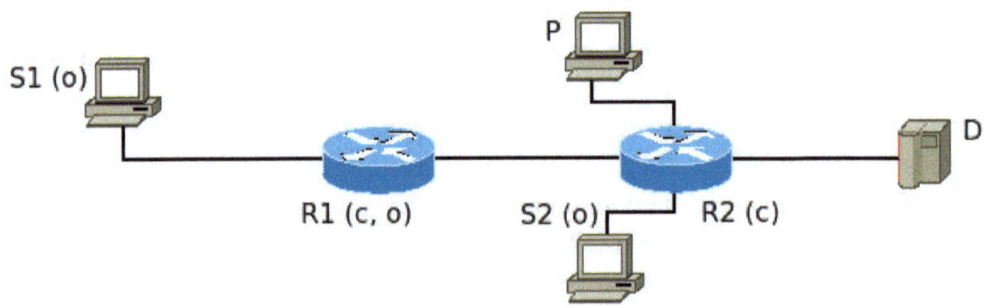

Fig. 1 The experimental network

R1 and R2 are IP routers. Every link connecting the nodes is an isolated 10 Mb/s Ethernet segment. We deployed our queuing discipline at nodes S1, S2, R1 and R2. Round trip time *RTT* of every circuit consisting of neighbour nodes was below 1 ms except the pair R2-D, where it was 50 ms. We assumed that the discretization period T is 10 ms, however from the results we obtained we concluded that it is in fact realised with +/- 1 ms irregularities. Taking this into account, while calculating the reference value according to the conditions formulated in [4], we assumed that $T = 11$ ms and finally obtained $X^D = 30$ kB.

Scenario

At $t \approx 1$ s S1 establishes a TCP connection with D. So does S2 at $t \approx 2$ s. Both connections are fed with data so that they are able to emit continuously at 10 Mb/s.

At t ≈ 4 s P starts sending data to D. It utilises UDP datagrams and does not perform any form of flow control. The configuration of R2 provides that the packets received from P take precedence over the ones received from S1 and S2. This causes the transmission of TCP flows by R2 to be suspended during the activity of P.

The scenario mentioned above was run in three variants. In variant 1 there was no additional (beyond the standard TCP congestion widow mechanism) flow control applied to any node. In variant 2 proposed flow control solution was applied to configurations R1-R2 and S2-R2, but not to S1-R1. This refers to situation when the considered network consists of core nodes that are managed by network administrator and leaf nodes that are managed by independent means (i.e. owners of mobile systems). Finally, in variant 3 flow control was also applied to pair of nodes S1-R1.

Experiment Results

Having in mind the aim of the experiment we concentrate on results that were acquired in nodes R1 and R2. There are two sources of output data. Whole network traffic that was transmitted by both R1 and R2 was recorded by *tcpdump* utility [17]. These records were processed using *wireshark* traffic analyser [18] to obtain the transmission rates of considered flows. Moreover, the logs, recorded by the queuing discipline every discretization period T, were utilised. Among others, such a log contains information on queue length (denoted as $x(t)$), amount of data that was lost during the last discretization period (denoted as $l(t)$; the unit of this value is B/T – bytes per period) and total amount of data that was lost during the experiment.

The results of the test run in variant 1 are shown on figures below. Fig. 2. depicts transmission rates measured in router R2, and Fig. 3. presents queue lengths and data loss in both routers.

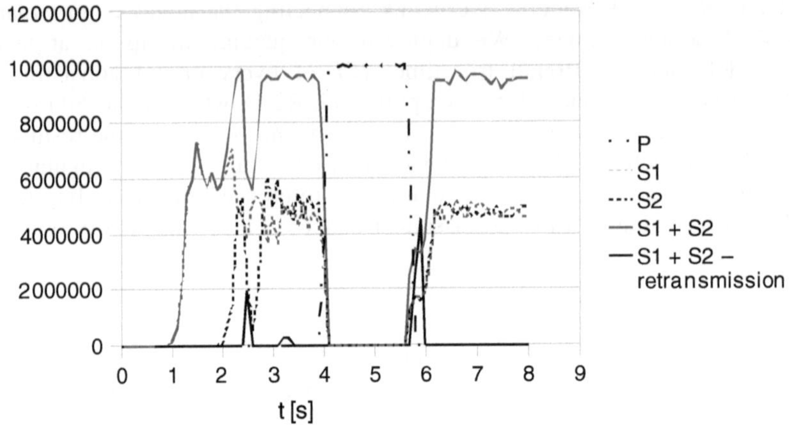

Fig. 2 Transmission rates, R2, variant 1

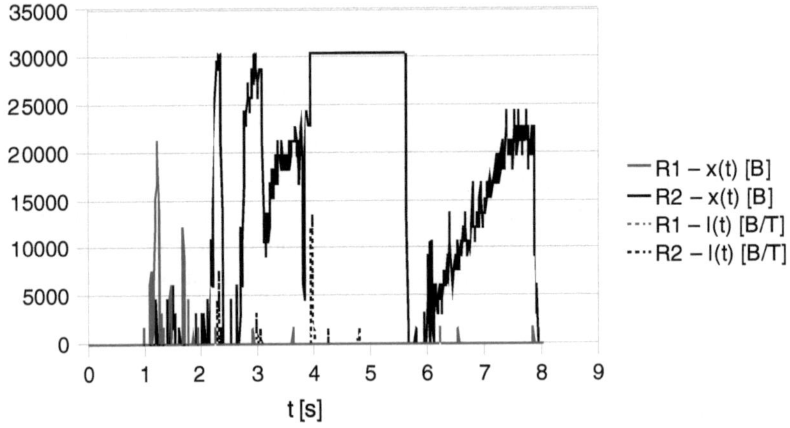

Fig. 3 Queue state, R1 and R2, variant 1

First we focus on the transmission rates of TCP connections between 2nd and 4th second of experiment. As TCP flow from S2 begins to transmit data, the flow from S1 is forced to reduce its congestion window and emission rate. Note that there is a perceptible amount of data being retransmitted during this period. Finally both TCP connections tend to equally share the throughput of link between R2 and D. One may observe similar results after node P finishes its transmission.

Let us concentrate now on the state of queuing disciplines in routers R1 and R2. The most noticeable phenomenon is that the average queue length in R1 is far lower than in case of R2. The reason is rather obvious: the available bandwidth of the link between R1 and R2 is the same as for the link between R1 and S1, thus the packets received by R1 are immediately passed to R2. This does not apply to the queue length of R2, especially during the activity of P, when the packets received from R1 and S2 are actually stopped and stored in the buffer. As the result congestion and data loss occurs in R2, especially at two moments: when S2 starts sending data and when P starts sending data.

Fig. 4. and Fig. 5. depict transmission rates measured in R2 and queue state observed in R1 in R2, respectively, during the test run in variant 2.

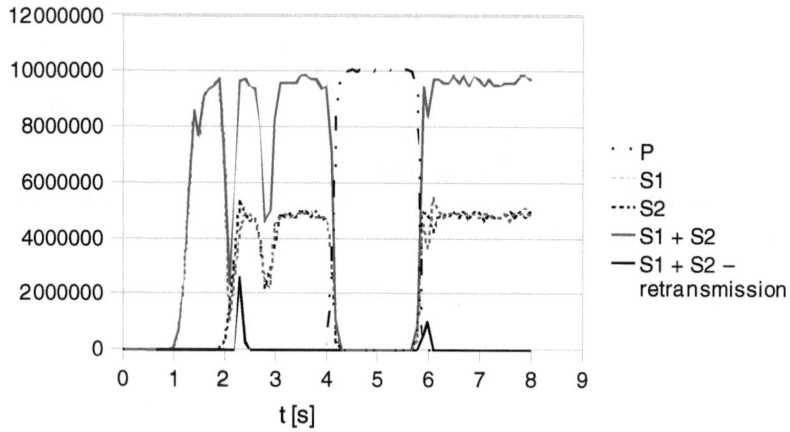

Fig. 4 Transmission rates, R2, variant 2

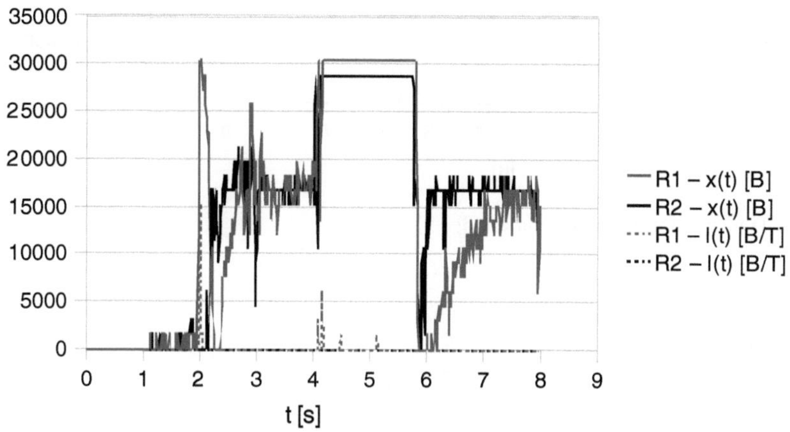

Fig. 5 Queue state, R1 and R2, variant 2

In this variant transmission rates of both TCP connection, observed from the perspective of R2, are limited by the proposed control algorithm that is applied to configurations R1-R2 and R2-S2 at network layer level. It is noteworthy that the available throughput is shared between TCP flows more smoothly and efficiently than in case of variant 1 (it is confirmed by the statistics presented at the end of the section). However, the most important changes (compared to results of variant 1) can be found through investigation on the state of packet queues in R1 and R2.

The first thing that is to be emphasized is that the congestion and data loss is completely eliminated for R2. Instead, we observe that these phenomenons occur in R1. This is a clear consequence of the fact that both R1 and S2 are controlled by R2 so that they do not send excessive amounts of data. Said that we have to consider a virtual discrepancy between the above remarks and the retransmissions (reported by *wireshark* analyser) that can be easily found on Fig. 4. We thoroughly inspected the traffic record obtained in R2 and concluded that there were no repetitive TCP segments found. From this we infer that the problem of retransmissions did not affect R2, as considered data was transmitted by this node once. This conclusion obviously does not concern R1 (namely, the link between S1 and R1), where data loss and retransmissions occur.

Finally, the results of variant 3 are presented in Fig. 6. (transmission rates in R2) and Fig. 7. (queue state in R1 and R2).

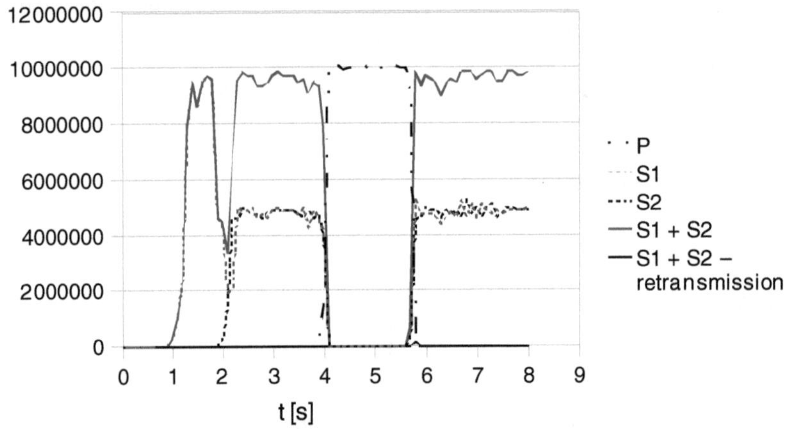

Fig. 6 Transmission rates, R2, variant 3

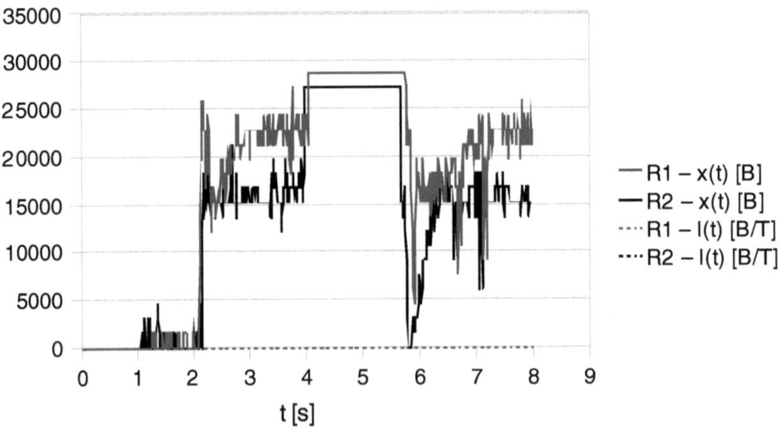

Fig. 7 Queue state, R1 and R2, variant 3

As the proposed control solution was applied to every considered pair of nodes, the problem of congestion, data loss and retransmissions was actually solved for both routers. However, the results obtained in this variant of the experiment show that there is at least on unfavourable effect of the control law we applied. Considering packet queue length on R1 and R2 depicted in Fig. 7. one can clearly observe that in case that there is no throughput available for TCP transmissions in R2, the queue length grows to its upper constraint not only in R2, but also in R1. This observation can be generalised as follows: if the throughput, that available for considered flows at the specified node, is reduced, the queue lengths in all the nodes preceding the specified one start to raise. As a consequence, the transmission delay of the packets, measured through the whole route from the source to the destination, enormously grows, what is obviously unwelcome.

In addition to the results presented above, total amount of data transmitted by both TCP connections through router R2 was measured in every variant of the scenario. Subtracting the amount of data that was retransmitted, we obtained so called "good throughput" – a effective transmission statistics.

Table 1 Amounts of data transmitted through router R2

Quantity [B]	Variant 1	Variant 2	Variant 3
S1 – transmission	3014936	3215984	3277850
S1 – retransmission	45420	52990	0
S2 – transmission	2204424	2376210	2464832
S2 – retransmission	36336	0	0
S1+S2 – transmission	5219360	5592194	5742682
S1+S2 – retransmission	81756	52990	0
S1+S2 – effective transmission	*5137604*	*5539204*	*5742682*

The results presented in Table 1. can be summarised as follows. The most important factor is the amount of data that was effectively transmitted to destination D, located in the last row. Taking as the base the result of experiment run in variant 1, one can easily see that applying the proposed solution within selected area of considered network (variant 2) caused that the good throughput was increased by 8.8%. Furthermore, if the solution was applied to every considered configuration, that growth was even greater (11.8%). The conclusion that must be drawn here is that the flow control algorithm designed by us significantly improves the efficiency of throughput utilization in the nodes that it is deployed at.

4 Summary

In this chapter a new flow control solution was examined. We focused at the situation when only a bounded area of the considered network was controlled by the proposed algorithm, what is related to a network with leaf nodes being administered autonomously (i.e. by the owners in case of mobile systems), not by the network administrator. The results that we obtained show that the desirable features of our control algorithm (such as congestion and data loss avoidance, or improved throughput utilization) are still valid within the controlled area. It is worth considering that it would not be achieved without making proper choices of: the ISO/OSI model layer that the solution is applied to (network layer), and the configuration of feedback closed loop (link-level control approach).

References

[1] Bartoszewicz, A.: Nonlinear flow control strategies for connection-oriented communication networks. IEEE Proceedings on Control Theory and Applications 153(1), 21–28 (2006)

[2] Bartoszewicz, A., Karbowańczyk, M.: Sampled time flow control algorithm for fast connection oriented communication networks. Journal of Applied Computer Science 11(1), 5–16 (2003)

[3] Chatranon, G., et al.: A survey of TCP-friendly router-based AQM schemes. Computer Communications 27(15), 1424–1440 (2004)

[4] Karbowańczyk, M.: Dyskretny algorytm sterowania ruchem w pojedynczym połączeniu połączeniowej sieci telekomunikacyjnej. In: XV Konferencja Sieci i Systemy Informatyczne – teoria, projekty, wdrożenia, aplikacje (2007)

[5] Qureshi, B., Othman, M., Hamid, N.A.W.: Progress in various TCP variants. In: 2nd International Conference on Computer, Control and Communication (2009)

[6] Ignaciuk, P., Bartoszewicz, A.: Linear quadratic optimal sliding mode flow control for connection-oriented communication networks. International Journal of Robust and Nonlinear Control 19, 442–461 (2009)

[7] Ignaciuk, P., Bartoszewicz, A.: Discrete-time linear-quadratic (LQ) optimal and nonlinear flow control in multi-source connection-oriented communication networks. European Transactions on Telecommunications 20, 679–688 (2009)

[8] Bartoszewicz, A., Molik, T., Ignaciuk, P.: Discrete time congestion controllers for multi-source connection-oriented communication networks. International Journal of Control 82(7), 1237–1252 (2009)

[9] Żuk, J., Bartoszewicz, A.: Discrete time sliding mode flow controller for multi-source single-bottleneck connection-oriented communication networks. Journal of Vibration and Control 15(11), 1745–1760 (2009)

[10] Katabi, D., et al.: Congestion Control for High Bandwidth-Delay Product Networks. In: Proceedings of ACM SIGCOMM (2002)

[11] Karbowańczyk, M., Duszyński, P., Lichy, K.: A simple, rate-based traffic control solution for IPv4 networks using Linux-based hop-by-hop closed loop and queuing discipline implementation. Information Systems Architecture and Technology: Service Oriented Distributed Systems: Concepts and Infrastructure, 313–325 (2009)

[12] Duszyński, P.: Implementacja pętli zwrotnej w sieciach IPv4 z wykorzystaniem mechanizmu śledzenia połączeń w jądrze Linux. M. Sc. Thesis, Technical University of Łódź, Institute of Information Technology, Division of Computer Networks (2009)

[13] Ha, S., Rhee, I., Xu, L.: CUBIC: A New TCP-Friendly High-Speed TCP Variant. ACM SIGOPS Operating System Review 42(5) (2008)

[14] `http://www.kernel.org/`

[15] `http://netcat.sourceforge.net/`

[16] `http://www.hping.org/`

[17] `http://www.tcpdump.org/`

[18] `http://www.wireshark.org/`

Planning of the 802.11/abgn Computer Networks with the Aid of the Specialized Web Service

Daniel Arendt, Michał Morawski, and Antoni Zajączkowski

Abstract. The chapter presents a systematic, computer aided method supporting network planning for interior wireless networks based on IEEE 802.11 standards family. Detailed specification of features and functionalities of the proposed network planning tool available as a web service is provided together with directions of its possible expansion. Description of the service is proceeded by critical evaluation of existing commercial and noncommercial software tools for network planning.

1 Introduction

Computer networks based on standards belonging to the group 802.11 are now very popular. Practically in every place people demand an access to the Internet. In common belief application of this network technology is very simple – it suffices to purchase an access point, configure its basic functions, place it in a suitable location and then benefit from its operation. Many laymen believe that this scenario will work. Unfortunately, everyday practice shows that in many cases a computer network created in an ad hoc manner is troublesome for its users, inefficient in operation and prone to disturbances. In addition one can encounter other important problems, not considered in our paper, like security and quality of communication[1] and mobility of workstations.

Several problems can be removed or minimized by correction of positions of computer stations and first of all positions of access points, selection of suitable antenna systems or a change of communication channels. Despite the fact that these actions are relatively easy, their effect is frequently unsatisfactory because in most cases they are performed using the trial and error method which in general is of poor efficiency.

Daniel Arendt · Michał Morawski · Antoni Zajączkowski
Technical University of Łódź, Institute of Information Technology
Wolczanska 115, 90-924 Lodz, Poland
e-mail: daniel.arendt@p.lodz.pl, michal.morawski@p.lodz.pl, antoni.zajaczkowski@p.lodz.pl

[1] The quality of communications should be considered as a provision of a maximum throughput for all users – in details lack or a very limited number of retransmissions, high bandwidth for all users.

P. Lipiński, K. Świrski (Eds.): Towards Modern Collaborative Knowl. Sharing Syst., SCI 401, pp. 157–171.
springerlink.com

In polish reality this last approach is particularly ineffective, because, as opposed to light and almost transparent interiors of office buildings in United States, in Poland dominate heavy buildings erected using materials of high attenuation ratio in the case of electromagnetic waves propagation. It is worth mentioning that even partition walls in many buildings in Poland are not constructed from the popular carton-gypsum cardboards but quite frequently from solid bricks.

For these reasons design and planning of wireless computer networks should be supported by some computational methods preferably available as a main part of a specialized web service. These methods on the basis of technical information concerning construction of the building and accessibility and efficiency specification of the network should provide means for reasonable placement of access points and antenna systems.

The organization of the paper is as follows. Section 2 presents basic formulae used by scientists and engineers for computation of the signal power attenuation. In Section 3 the most widely known network planning software tools are described. The next Section presents our approach to the design of an effective and accessible web service for network planning. In Section 5 some conclusions are formulated and further research directions are indicated.

2 Planning Wireless Computer Networks – Computations

Detailed discussion of computational methods applied for design of antenna systems can be found for example in [16]. Hence, we confine discussion of these methods to basic relations relevant for our further considerations.

In the case of most simple dipoles the power propagated in the dielectric medium is given by the formula

$$P(r) = \sqrt{\frac{\mu}{\varepsilon}} \left(\frac{I_0 l}{2\lambda r} \right)^2 \sin^2(\theta) \cos^2(\omega t - \beta r)\,, \tag{1}$$

where μ and ε denote the magnetic and electric constants of the medium, λ denotes the wave length, l denotes the dipole length, θ denotes the angle between the axis connecting radiating elements and the direction of radiation, β denotes the phase shift, r denotes the distance between the dipole and the given point and I_0 denotes the antenna current.

This formula is sufficiently accurate when the condition $r >> \lambda$ is satisfied – the far zone. It is easy to notice that this formula despite its theoretical correctness has limited applicability in engineering practice of network planning. This comes from the fact that values of some parameters are hardly available, as in the case of the antenna current. For other types of antenna the respective formulae are even more complex, so engineers concerned with network planning prefer more simpler but in many cases more useful formula expressing the power loss in the form

$$P_\lambda(r) = k\lambda^2 r^2\,, \tag{2}$$

where the coefficient k depends on the medium and it generally depends on the frequency see Table 1, but in the particular band is considered as a constant.

If an antenna is of the directional type, then

$$P_\lambda(r) = k\lambda^n r^n \,,\; n \in (1,2)\,. \tag{3}$$

In practice the exponent in formula (3) (n) for a given construction of a store, flat, or office equipped with furniture can vary up to 5. Moreover, some media which should be considered cannot be classified as dielectrics. To this class belong steel structural components such as concrete reinforcements or lift shafts but also reinforced window panels, window panels coated by the thin metal layer and some kinds of plaster. In such cases precise analysis is much more complex and respective material parameters are non-linear and frequently uncertain. Hence, in engineering practice different approximate models are used. These models differ according to the particular type of the network considered – interior or exterior.

Because here we are interested in interior wireless networks the following attenuation formulae valid for the air medium are used [14, chap 7]

$$P_{2.4Ghz} = 55 + 0.3r[dB]\,, \tag{4}$$

and

$$P_{5.5Ghz} = 63 + 0.3r[dB]\,, \tag{5}$$

where r denotes the distance between a transmitter and a receiver.

These formulae are valid for $r > 3, m$ and we consider it as a reasonable assumption for our further considerations. The formulae (4) and (5) are very simplified in order to achieve the efficiency of computations. In the subsequent work we plan to find the proper balance between precision of calculations and the resource consumption based on [2, 5, 9, 20].

In the case of other materials – not air, the similar formula is applied

$$P(r) = k_p + k_l r_m\,, \tag{6}$$

where the coefficients k_p and k_l depends on the kind of material and the transmission frequency, r_m is the length of path in this material.

Due to the fact that companies engaged in commercial design of computer networks do not publish precise values of this coefficient its values are not widely known and only average values are easily available [1].

Typical programs, described in the next Section, compute and present only the radiation power distribution of the access points provided that there are no reflections. They do not take into account interaction of workstations which results in neglecting the problem of the hidden workstations.

This feature of existing software tools supporting network planning can be regarded as its basic weakness and therefore we consider in our research to include the analysis of signal reflection into our network planning tool, of course, at the expense of its increased computational complexity. In our opinion this functionality could be particularly useful in the case of the network standard 802.11n. The selection of the propagation model appropriate for such situation will be performed after first experience gained from the usage of the developed service.

Approximate values of radio waves attenuation for the most popular materials and components encountered in buildings are given in Table 1.

Because there is a strong correlation between the power of the received signal and the transmission bit-rate, a correct placement of active elements of communication system is of prime importance for increase of the communication efficiency. Even only one station placed in the domain of bad conditions can substantially diminish operation efficiency of the whole network, and as a consequence of these stations which are placed in domains of satisfactory conditions.

Table 1 Radio wave attenuation for selected materials and components

Material/Component	*5GHz [dB]*	*2.4GHz [dB]*
Solid wood door (1.75”)	10	6
Hollow wood door (1.75”)	7	4
Interior office door with glass panel (1.75”/0.5”)	6	4
Steel Fire/Exit Door (1.75”)	25	13
Steel Fire/Exit Door (2.5”)	32	19
Steel Rollup Door (1.5”)	19	11
Brick (3.5”)	10	6
Concrete wall (18”)	30	18
Cubical wall (fabric) 2.25"	30	18
Exterior concrete wall (27")	45	53
Glass divider (0.5")	8	12
Interior hollow wall (4")	3	5
Interior hollow wall (6")	4	9
Interior solid wall (5")	16	14
Marble (2")	10	6
Bullet-proof glass (1")	20	10
Exterior double pane coated glass (1")	20	13
Exterior single pane window (0.5")	6	7
Interior office window (1")	6	3
Safety glass-wire (0.25”)	2	3
Safety glass-wire (1.0”)	18	13

3 Planning the System of Access Points – Existing Software Support

Currently we can use several different software tools supporting planning and implementation of wireless networks.

A complete solution is offered for the case of exterior networks and is known as Radio Mobile. However, this tool has its main drawback - the inconvenient, manual way of entering data required for execution of desired computations.

Similar features offer other programs, described in the next Subsections. Each of these programs includes a module for design of exterior networks operating in areas of diversified topography, but as was already mentioned, we are rather interested in design of the web service for interior networks planning with possibility of entering large amounts of data describing construction of the building according to the standard formats while keeping the data volume at the reasonable minimum.

In subsequent four Subsections we shortly discuss main features of most popular programs for computer aided planning of interior networks.

3.1 Air Magnet – The Planner Service

One of the best known and probably one of the first network planning tools is the system AirMagnet Planner dedicated for wireless networks. Using construction data of the building an operation of the network is simulated. Additional data concerning characteristics of the network hardware are needed: power, frequency and antenna characteristics. As a result a power covering map of the building is obtained with marked areas of excessive interference. A very important feature of the program is possibility of importing a map of the particular object provided in a number of standard formats including DWG directly from AutoCAD™. Unfortunately it does not suffice to import a plan of the building, because using a specialized tool we need to mark positions, where some special elements of equipment like doors, furniture, windows etc. are placed and then we need to specify material characteristics of these elements.

As a result of program operation we obtain a colorful map of the signal power, similar to that shown in Figure 1.

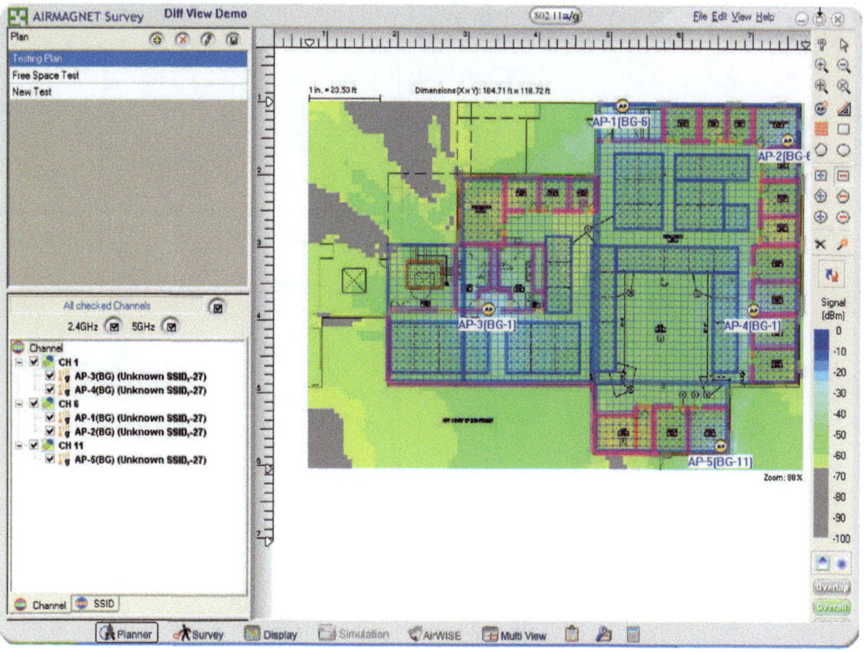

Fig. 1 An exemplary map of the signal power obtained for some exterior area [8]

Functionality of AirMagnet Planner can be enhanced by the self-acting module for the access points placement (Planner Advisor) which needs some additional specification of parameters concerning requirements of covering the object by the network signal.

3.2 HeatMapper and Site Survey of Ekahau

One of the programs for wireless networks planning in the interior of a building is Ekahau HeatMapper [3, 4]. This software is dedicated first of all for recording signal power of the existing network and provides a power image useful for design, analysis and recording of the wireless network operation.

For satisfactory operation of this program we need a mobile computer with the network interface and in order to calibrate measurements we need also an outline of the room, floor or building to be investigated. It is worth mentioning that the program can also be used in an open space.

Additionally, the program enables to choose a group of workstation recorded. The group can be composed of all detected access points or the group can be confined to access points with the selected parameter like SSID, channel number or other.

In order to collect data we should explore a particular area using a measurement set including a computer with a wireless network card and the program running. During the trip we input our subsequent positions by clicking a mouse on respective points on the application window. The program produces a map of signal power as shown in Figure 2.

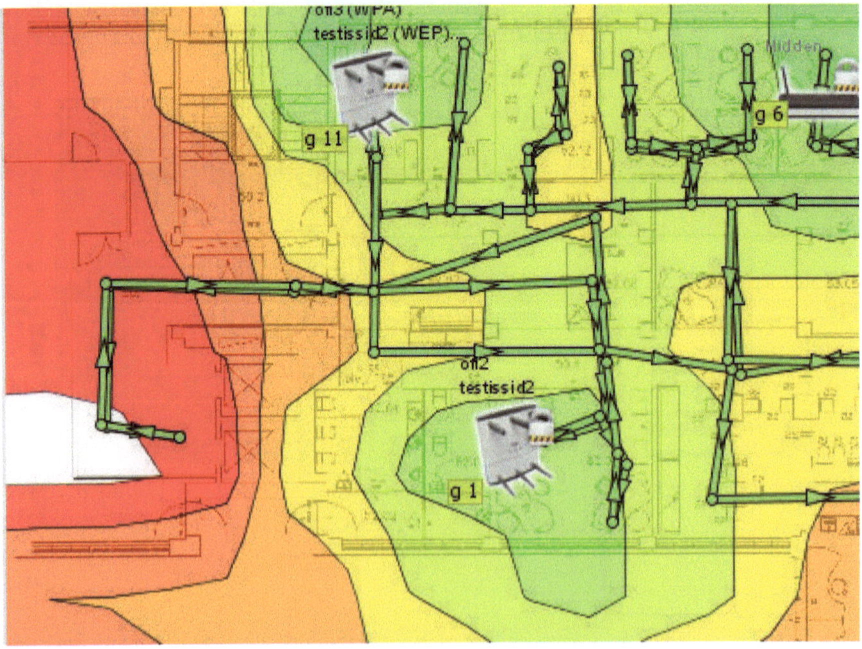

Fig. 2 An exemplary map of the signal power obtained for some area [3].

On the map we obtain a path (green line) passed by the measurement set and the colorful image of the network signal power. Accuracy of the image depends on scrupulosity of the person operating the program. Low speed of movement, frequent entering of the actual position and preparation of sufficiently dense grid of measurements are needed to obtain an image with negligible approximation and interpolation errors.

Ekahau HeatMapper belongs to the free-software category, but posses one serious disadvantage - the time of the measurement trip is limited to 15 minutes. From that follows that this program can be regarded rather as a demonstration of capabilities of Ekahau software then as a fully useful tool for wireless network planning.

Ekahau Site Survey is a commercial version of HeatMapper and is devoid of limitations of its free counterpart. The basic version has similar functionality as HeatMapper but with no time limit. The Professional version enables network planning on the basis of the building construction, or network planning in the three-dimensional space and is equipped with a number of useful tools as for example co-operation with a GPS receiver for automatic recording of the measurement trip on the calibrated map or an expert module supporting network expansion.

Similarly as in the case of other such programs one can obtain a "temperature" map of the signal power on the selected floor of the building as can be seen on Figure 3.

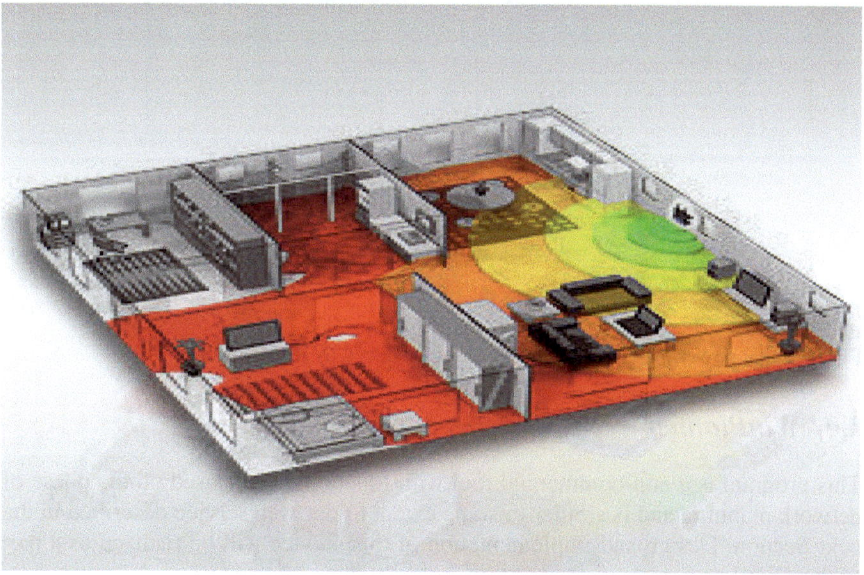

Fig. 3 Signal power covering of the building floor produced by Ekahau Site Survey [4]

3.3 RingMaster of TrapezeNetworks

The program RingMaster supports management of the whole life-cycle of the computer network and posses similar capabilities as other such programs [13]. By the life-cycle of the network we mean the following steps: initial planning, simulation, process of documentation, maintenance, diagnostics and modernization planning. From the point of view of network planning the most interesting feature of RingMaster is a module for planning and scaling which can import data concerning construction details of building provided in DWG and DXF formats and then integrates these data. This feature allows to minimize the time spent on arduous and troublesome process of introducing data required for network planning [6].

As a result a colorful map of the signal power is obtained for the given floor of the building as shown Figure 4. In [7] one can find prices for different configurations of RingMaster.

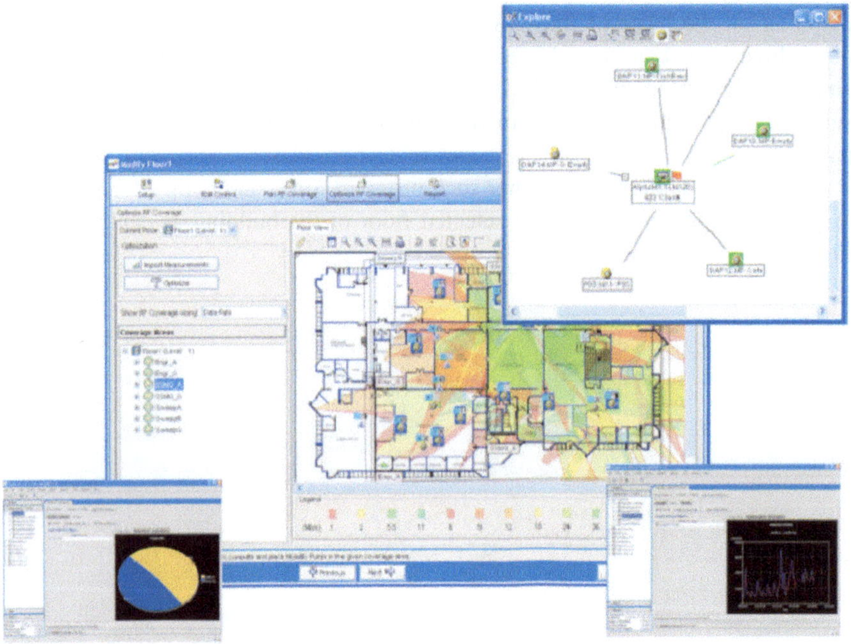

Fig. 4 Signal power map of the building floor produced by RingMaster.

3.4 WinPlanner

This program is a non-commercial tool with functionality focused on the phase of network planning and is similar to some extent to our web service described in the next Section. Design and implementation of this service will be realized as a part of the TEWI project.

WinPlanner has several interesting features like possibility of introduction the building plan prepared in AutoCAD. This feature illustrates Figure 5.

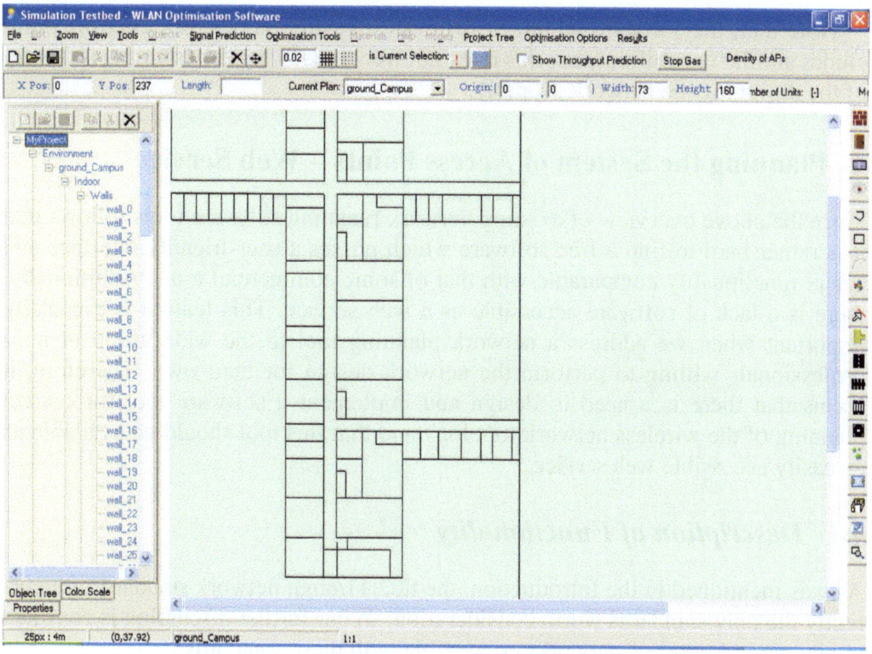

Fig. 5 WinPlanner screen view with the plan of the building floor [18].

However, authors of WinPlanner concentrated most of their effort on implementation of automatic localization of access points. This functionality is almost self-acting after introduction of only just several parameters: the maximal number of access points, the number of employees for the surface unit and the power level. The program does not use information concerning wall materials and their properties. Additionally it is assumed that all antennas are unidirectional and that the antenna output power is the same in all base stations. From these follows that this extraordinary interesting and scientifically advanced program [11, 18] has limited capabilities as a computational tool applied in practice for network planning. In our opinion the program can be rather considered as an interesting result of scientific research then a fully useful engineering tool. The program produces the signal power map similar to those described in former Subsections.

3.5 Other Solutions

The solutions mentioned above do not fulfill the set of available tools for network planning, but in our opinion they are most widespread. However, it is certainly worth to mention VisiWave [17], RF3D [15], LANPlanner [12] or layer-1 tools like e.g. Wi-Spy [19], and many others. None of them has unique capabilities so they are not described in this paper.

Note that, the software inbuilt in firmware of wireless network controllers includes similar capabilities, but this does not free the developers from the necessity of the designing the network properly.

4 Planning the System of Access Points – Web Service

From the above overview of existing network planning software tools follows that it is rather hard to find a free software which posses a user-friendly interface and offers functionality comparable with that of some commercial tools. Additionally, there is a lack of software accessible as a web service. This feature is especially important when we address a network planning tool to the wide group of non-professionals willing to perform the network design for their own. Therefore, it seems that there is a need to design and implement a software tool for correct planning of the wireless network topology and that this tool should be available as an easily accessible web service.

4.1 Description of Functionality

As was mentioned in the Introduction, the 802.11/abgn network standards are currently the only standards worth consideration, so our further discussion is confined to planning of wireless networks compatible with these standards.

Another assumption is that our network planning tool should be implemented as a web service with widely accepted MVC multilayer structure composed of the data-base server, application server responsible for computations and interpretation of user requests and web server for realization of the user interface.

We have decided to implement in our web service the following basic functions:

- Handling individual user accounts
- On the user side access by standard WWW browsers
- Service should comprise a data base of materials and their energy attenuation parameters for 2.4 GHz and 5.5 GHz bandwidths. This data base should also posses other useful properties and contain additional information like:
  - Initial information provided by the data administrator – the global part
  - Dictionary supporting physical integration of identical materials described in most popular European languages
  - Dictionary supporting physical integration of similar materials described by different names in the same language
  - Individual data base of materials attached to the particular user - the user part
  - Ability for controlled sharing of material information
  - Information on national regulations in EU countries concerning power and frequency of radio devices operating in the 802.11/abgn bandwidths and admitted for use without the separate permission

- Data base should include initially provided antenna characteristics and should enable to define the new ones. It should also contain the global and user parts, but it seems that language versions are of little concern.
- Service should accept description of the building construction in formats typically used for design of buildings and their parts
- Service should allow for placement of wireless access points and antennas on the plan of the building and supplement this information by other relevant operational conditions
- Service should provide means for specification of the following expected boundary values of parameters acceptable during the particular network design:
  - Maximal admissible power (possibly higher then the one specified by national standards)
  - Minimal expected power – related to the predicted and expected data flow.

The plan of an object and specification of access points form the input data for the service. After introduction of these data the application server can start network planning computations. As a result a person using our service should obtain the colorful map of signal power similar to the well-known temperature or geographical maps. On this image areas with excessive values of selected design parameters and areas of potentially poor network operation (channel collision, etc.) should be visibly identified.

We also assume some other features of the service. We do not expect that computations will be always performed in the real-time as seen by the client. On the contrary, we allow operation like in the batch mode. To be more precise, a client can introduce and leave her/his input data and when the application server completes network planning the web server informs a client via an e-mail or sends results as an e-mail attachment. This would be especially useful when more complex propagation models will be used.

4.2 Functional Diagram and Information Flow

On the basis of functional description of the proposed network planning tool we can build a general block diagram of the service. This diagram, presented in Figure 6, shows main modules, explains their roles and operation and an information exchange. As was mentioned in former Sections, clients use standard web browsers in order to access the service, to input their own data and to view results of computations. The service is designed according to the classical model-view-controller architecture. In the data base we can find three logically disjoint parts. The first part, governed by the service administrator contains information concerning limitations imposed by national standards on wireless network construction, antenna characteristics library, material characteristics library together with the respective dictionary and exemplary design patterns. The second part belongs to the clients which can there input construction plans of buildings or floors, specify domains of network accessibility and provide individual requirements and constraints. Clients can define new materials and their parameters or modify parameters of globally defined materials by masking the old values by the new ones. The third part constitute data bound with operation of the application server and information processing logic.

This part is also devoted for exchange of information collected during experiments and measurements performed on the real network. It means that results of computations performed by the application server are confronted with the results of measurements.

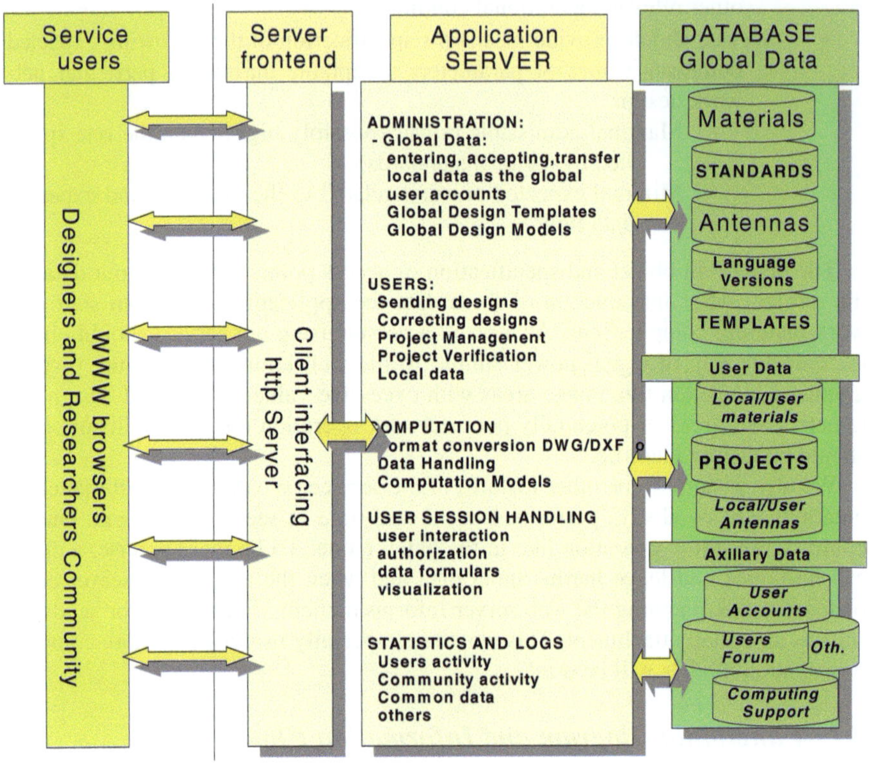

Fig. 6 The block diagram of the network planning service with main modules and information flow.

We expect that the proposed approach to implementation of network planning tools can be an interesting solution for ever growing community of designers, engineers and researchers involved in network design and planning.

4.3 Life Cycle of the Web Service

After starting the service an administrator must fill in the first part of the data base by the global information including wireless network standards, antenna characteristics and basic parameters of the most popular building materials. It is assumed that clients providing data for their own projects will supplement material data and correct some values of material parameters such as energy attenuation in bandwidths reserved for wireless networks compatible with 802.11/abgn standards. The service administrator must assess quality of the data provided by the users and in the case of positive assessment should transfer these data from the user part to the global part of the data base. The process of data validation can be supported by opinions and discussion with service clients. We expect that in this manner the attenuation characteristics of building components and construction materials will be far more accurate and extensive as those given in Table 1.

4.4 Next Version of the Network Planning Web Service

After successful implementation and testing of the proposed network planning tool we expect its further development by expansion of its functionality with the aid of several new services exploiting in part the formerly collected data. The new services include among other things shared data bases of material properties, antenna characteristics and national standards with language and dictionary bindings and what is possibly more attractive an intelligent advisor responsible for automatic selection of access points positions. In other words, this functionality can be regarded as a kind of virtual client providing repeatedly a number of different access point distributions on an orthogonal grid with density selected according to technical requirements used in network planning – for example 0.5 m. The virtual client systematically moves access points and after each iteration of computations validates the resulting access point distribution as compared with former ones. This feature of the network planning tool can be encountered in some programs discussed in Section 3 and is an area of scientific investigations [10, 11]. At the current stage of our research it is rather hard to point out methods supporting automatic planning of the access point system. This aspect of our research will be discussed in a separate paper.

5 Conclusions and Further Research

Planning of wireless local area networks compatible with IEEE 802.11 standards is currently investigated by many computer network specialists. Several software tools supporting the process of network planning are offered, but in many cases their application is troublesome due to the fact that a lot of data concerning building construction, access points characteristics or material attenuation coefficients must be provided manually. Therefore it seems to us that a network planning tool in the form of an open-source program accessible via a typical web browser can be an attractive solution for many persons engaged in network planning. The proposed planning tool possesses an important feature of strongly limited amount of initial data needed for the effective operation of the tool. As a result a client should obtain an access points distribution which is a vital part of an economical and robust wireless network. This is possible due to requirement that building construction details are introduced in standard formats used by popular CAD programs. This approach is not unique but is not widely used. However, the more important and interesting feature of our network planning service lies in its data base organization and handling. Firstly, data concerning national standards of admissible power levels EIRP propagated by access points are provided only once and secondly, the data base containing values of material parameters and attenuation coefficients for bandwidths of IEEE 802.11 standards is minimized at the start of service operation and then it is expanded by the community of engineers and designers using our service.

Further research will be concentrated on methods for automatic planning of the access point system and in parallel on software implementation of our network planning service.

References

[1] 3COM Wireless Antennas Product Guide, http://www.3com.com/other/pdfs/products/en_US/101900.pdf

[2] Abhayawardhana, V.S., Wassell, I.J., Crosby, D., Sellars, M.P., Brown, M.G.: Comparison of Empirical Propagation Path Loss Models for Fixed Wireless Access Systems. In: 2005 IEEE 61st Vehicular Technology Conference, VTC 2005. Spring (2005), http://www.cl.cam.ac.uk/research/dtg/lcepub/public/vsa23/VTC05_Empirical.pdf

[3] Description of Ekahau HeatMapper, http://www.ekahau.com/products/heatmapper/how-to-use.html

[4] Versions of Ekahau, http://www.ekahau.com/products/heatmapper/feature-comparison.html

[5] Domazetovic, A., Greenstein, L.J., Mandayam, N.B., Seskar, I.: Propagation Models for Short-RangeWireless Channels with Predictable Path Geometries. Wireless Information Network Laboratory (WINLAB), Rutgers University, http://www.winlab.rutgers.edu/~narayan/PAPERS/DGMS_Trans.pdf

[6] Description of Trapeze RingMaster, http://www.trapezenetworks.com/products/ringmaster_software/

[7] Price list of Trapeze RingMaster, http://www.peppm.org/Products/Trapeze/price.pdf

[8] Description of AIRMagnet, http://www.airmagnet.com

[9] Frenkiel, R.H., Badrinath, B.R., Borras, J., Yates, R.D.: The Infostations Challenge: Balancing Cost and Ubiquity in Delivering Wireless Data. IEEE Personal Communications 7(2), 66–71 (2000)

[10] Gibney, A.M., Klepal, M., Pesch, D.: Evaluation of developed WLAN Optimization Tool. In: Proc. CSSN IT&T Conference, Carlow, Ireland (October 2006)

[11] Gibney, A.M., Klepal, M., Pesch, D.: A Wireless Local Area Network Modeling Tool for Scalable Indoor Access Point Placement Optimisation. In: Annual Simulation Symposium, ANSS 2010, Orlando, Florida, USA (April 2010)

[12] LANPlanner, http://www.motorola.com/Business/US-EN/Business+Product+and+Services/Software+and+Applications/Network+Design+Software/LANPlanner_US-EN

[13] Oppenheimer, P.: Network design using the top-down approach. PWN, Warsaw (2007) (in Polish)
[14] Orthman, F., Roeder, K.: Wi-Fi Handbook: building 802.11b wireless networks. McGraw-Hill (2004)
[15] RF3D WiFi Planner 2, `http://emea.psiberdata.com/en/home/products/by-application/wifi-simulation/rf3d-wifiplanner.html`
[16] Rosłoniec, S.: Fundamentals of antenna technology. Warsaw University of Technology Publishers, Warsaw (2006) (in Polish)
[17] VisiWave Site Survey, `http://visiwave.com`
[18] WinPlanner, WLAN Optimisation Software v0.1, User's Guide v0.01, Ubiaware Technologies, Bishopstown, Cork, Ireland, `http://www.aws.cit.ie/ubiaware/docs/winplanner.pdf`
[19] Wi-Spy, `http://www.metageek.net/`
[20] Yates, R.D., Mandayam, N.B.: Challenges in Low-Cost Wireless Data Transmission. IEEE Signal Processing Magazine 17(3), 93–102 (2000)

Universal Snapshot File Concept

Mateusz Smoliński

Abstract. Snapshots are often used to perform on-line data backup, when storage is working without downtime. Nowadays, snapshots functionality is offered by several filesystems such ZFS, BTRFS or it can be achieved independently from the filesystem utilized, using block-device layer i.e. LVM. Existing snapshot is related to selected block device or filesystem technology, so it is not portable, especially when considered data are migrated to other filesystem. Concept of universal file snapshot is not limited by concrete filesystem, but it can be used in various file system. Universal snapshot file is based on the assumption that snapshot data are stored in special file. A version file that is related with considered universal snapshot file uses compensation process to provide access to previous version of original file data. In order to prevent the original file from corruption, all data modifications are separated by transactions, that use a journal file. Many other functionalities such as original file data history (local versioning) or point in time recovery utilize the universal snapshot file. The presented solution requires no user interaction. When one of original files is modified, all the necessary activities can be executed automatically. Moreover, universal snapshot file technology is transparent to the software that performs operations on selected original file.

1 Introduction

Modern filesystems such as ZFS (Zettabyte File System) or BTRFS (B-tree File System) introduce new functionality – a snapshot [1,2]. Taking into account a general filesystem, snapshot can be concerned with a single file or directory, which groups many files. Single filesystem original object may be connected with many snapshots. Every filesystem snapshot uses available filesystem resources to store its data.

Snapshot functionality is also offered at block device level i.e. LVM (Linux Volume Manager) [3]. This snapshot is transparent to filesystem, which uses block

Mateusz Smoliński
Technical University of Łódź, Institute of Information Technology
Wolczanska 115, 90-924 Lodz, Poland
e-mail: mateusz.smolinski@p.lodz.pl

P. Lipiński, K. Świrski (Eds.): Towards Modern Collaborative Knowl. Sharing Syst., SCI 401, pp. 173–183.
springerlink.com

device. Single snapshot represents data state of every block at the time, when snapshot was created. Block device level snapshot allocate additional blocks on other block device to store its data. Similarly to filesystem snapshots, one block device can be related to many snapshots, that represents different data states.

Both techniques allow system administrator to create read-only and read-write snapshots. Read-write snapshot utilizes a clone of data, which can be modified. Logically, data clone may be observed as another data instance, but physically it is not a separate copy of data (storage space occupied by data clone is usually smaller). Therefore filesystem based solutions can offer directory or file history management, but this versioning depends on dedicated software tools. User must be aware on using this software to preserve the history of owned files. Such a solution may obviously be effective only for files that are stored in concrete filesystem. Simple move operation to other filesystem (i.e. on pen-drive or network storage) causes that the versions history is lost and subsequent file data modifications can not be registered.

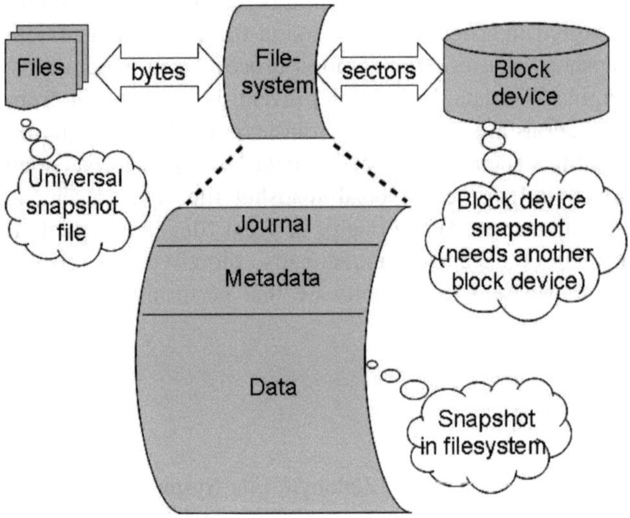

Fig. 1 Various level of snapshot technology

A new solution, which uses universal snapshot file, is devoid of presented disadvantages (figure. 1). It can be used in every filesystem. Also there is no need to use any dedicated tools, because operating system executes every indispensable operation. Every file modification immediately appears as a new file, preserving preceding data version from original file. This new file is not a copy, because presented solution uses file snapshots. It also uses transactions that groups many file modifications.

2 Transactional File Modifications

Nowadays filesystems utilize journals to register performed transactions. Transactions group filesystem operations on metadata like: resource reservations (i.e. blocks or i-nodes), data modification in reserved blocks and filesystem directory hierarchy changes. Single filesystem transaction does not always assure grouping of many simultaneous file modifications, because filesystem journal has fixed size (i.e. ext2/ext3). Therefore, even in cause of modern transactional filesystems, a file data loss may take place when an error occurs during writing process. Usually filesystem transactions assure only structures (metadata) consistency, so file data may be only partially updated. Furthermore, filesystem transactions (according to ACID: Atomicity, Consistency, Isolation and Durability [4]) are never rollback. The software that performs the operation may be able to recognize partial data change and report file corruption. Some of applications could even repair corrupted data (i.e. office tools: word processors or calculation sheets). In many cases a file containing incomplete data is useless at all. Generally there is no universal tool provided to restore previous version of data from corrupted file, because it is not stored in filesystem.

Universal snapshot file concept introduces a transactional file modifications using journal file. These transactions will be called logical to distinguish them from internal filesystem transactions. The journal file stores only records related to original file data modifications including add, modify and delete operations. A single record of journal file consists of:

- offset (address of data in file);
- data size;
- data;
- checksum including data and offset.

Journal file records and original file data are separated, so even if an error in write process occurs, there is no possibility to partial new data existence in original file. When logical transaction commits, which requires no error in data writing process, a synchronization process starts (figure 2). As a result of synchronization process data are copied from file journal to original file, and at that time no other original file modifications are permitted. Even when in this process encounters an error, no data will be lost and synchronization process can be finished later. After synchronization process finishes its work, logical transaction journal file is removed.

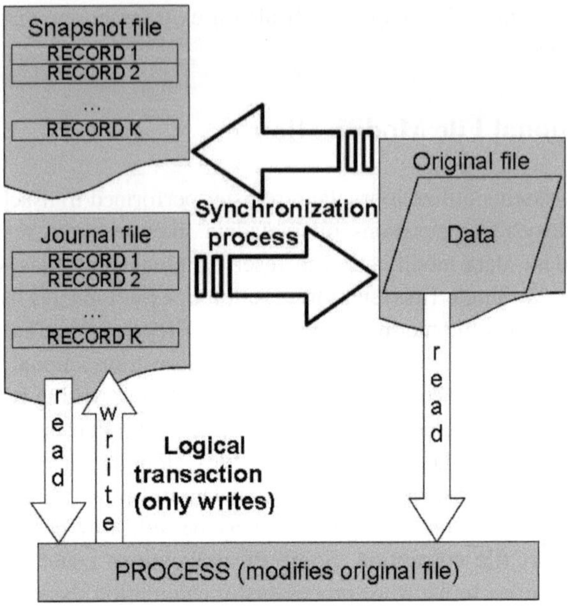

Fig. 2 Synchronization process

Logical transaction boundaries for many simultaneously file modifications could be established explicitly by special write (i.e. offset 0x0 with zero data size) or implicitly by fixed time limit that determines maximum period of time between two sequential file modifications belonging to single transaction. If there is no file modification after fixed time limit, current logical transaction is committed automatically. Any subsequent file modification begins a new logical transaction. Auto-committed transactions are transparent to external software, but they do not allow using long lived logical transactions.

If a filesystems provides advisory locks, many processes may modify file data simultaneously. In that case, to prevent logical transactions concurrency problems, only one from parallel executed transactions on single original file can be committed. Optimistic lock concept is based on original file data checksum (including time stamp of last file modification) as a version field that is examined before synchronization process begins its work. This is to solve concurrency problem between logical transactions that simultaneously work on the same original file. In this case only one logical transaction is permitted to commit and other transactions must be rolled back, because checksum does not match in version verification. Even logical transaction is rolled back, its journal file is not removed and it still stores all data modifications executed in this transaction. To prevent data lost each journal file related with rolled back logical transaction must be removed explicitly by its owner.

3 Related Snapshot and Version Files

Universal snapshot file is separated from original file and is created in synchronization process. According to the rule of COW (Copy-On-Write [5]), before data modification in original file occurs, modified data are copied to snapshot file. Modified data include also deleted data from original file. It is worth to mention that only the first modification of selected data range in logical transaction boundaries results in copying data to snapshot file. Therefore snapshot technique is not efficient when all data are replaced in file.

Snapshot file uses the same record format as journal file (presented in transactional file modifications) to store data. Every modification of specific data range requires adding one record to snapshot file. It is also required by the operation of adding new data to original file. All records must be written in snapshot file when synchronization process finishes its work, and then the dedicated version file is also created. Created version file is intended to be read only and its creation-time is identical with the same parameter of snapshot file.

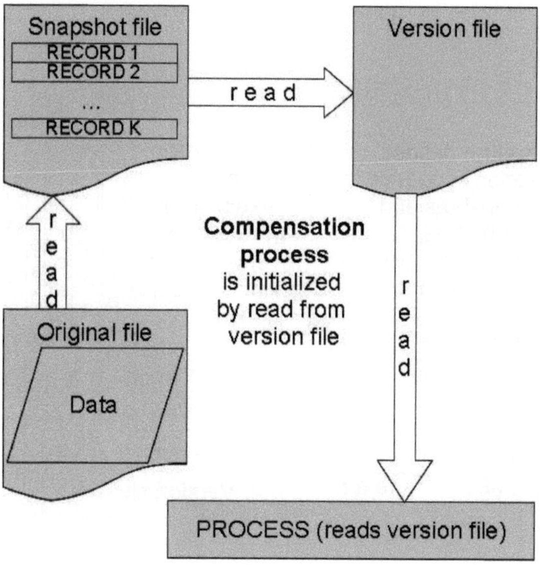

Fig. 3 Compensation process

Records stored in snapshot file are used by compensation process to provide access to previous versions of data from original file (figure 3). Therefore all record offset must be calculated to original file data after modifications. An execution of compensation process utilizes both data sources: snapshot file and original file. Compensation process is executed each time, when dedicated version file is read by external software. According to compensation process design, when requested data in read operation from version file are not stored in records of snapshot file, then they are read form original file. From logical point of view (i.e. for processes), version file stores previous version of data from original file, which size is declared as version file size.

4 File Data History

Single version file, using compensation process, provides access to previous version of data from original file. In this process records from single snapshot file are read. When logical transactions sequence is processed on original file and each synchronization process creates its own snapshot and version files, only previous version of data from original file is directly accessible. This is a result of fact, that every committed logical transaction creates snapshot file data base on current original file data. Nevertheless, using compensation process recursively it is possible to indirectly access any version of original file data related with one from sequence snapshot file. Snapshot files creation sequence is indispensable to keep original file data history (figure 4).

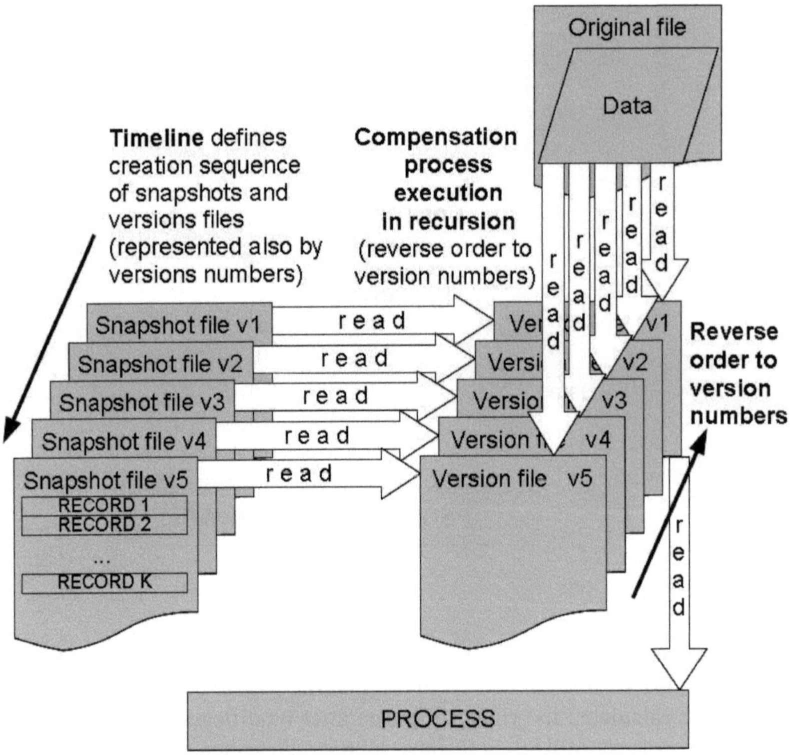

Fig. 4 File history created by version files

When the same data in one file are modified by different logical transactions, it is necessary to know which snapshot file was created earlier. Identical order is for version files creation. Order of snapshot file creation can be kept using version number. Newer snapshot file must be distinguishable from the earlier ones by greater version number. Therefore version number starts from one and could be add to file name in fixed format. Identical rule has to be applied to the name of the assigned version file, so that the related snapshot and version file have the same version number in its name.

User can always remove versions file, deleting with it related snapshot file, but deletion of snapshot file with version *N* blocks reading versions of data from earlier versions *M* of this file, where $M<N$. It is obviously implied by the fact that no records are duplicated in snapshot files. Without records from version *N* snapshot file, earlier records from snapshots file, which version $M<N$, are useless. Lack of record duplication in snapshot files assures efficient use of storage space required to keep file data history. It also reduces number of operations according to COW rule in transactional file modifications.

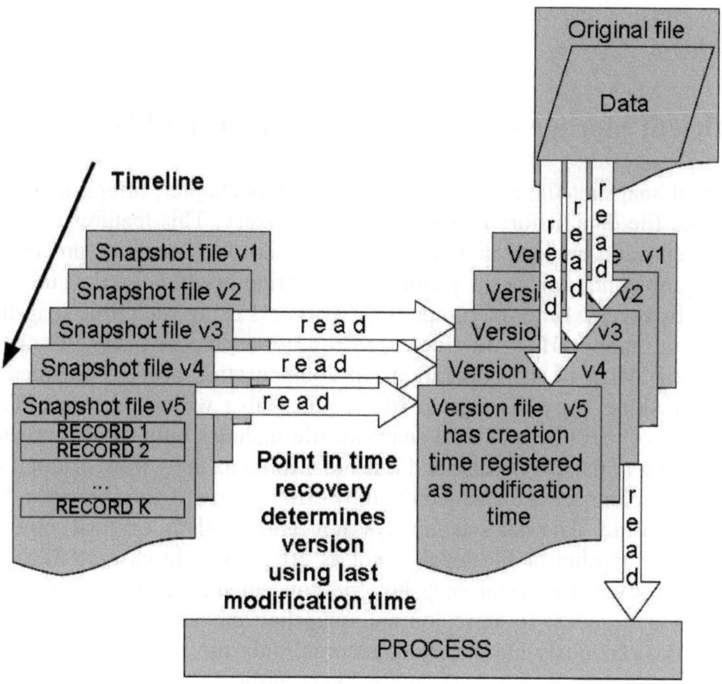

Fig. 5 Point in time recovery of original file data

Each snapshot and version file, created when logical transaction is committed, has creation time for executed original data modifications. The file creation time could be changed in simple file copy operation. When a file is read only, its modification time is always identical with creation time, so version file modification time determines finish of related logical transaction. Additionally, this could be verified by version number, when previous or next version file is present. This allows user to determine the moment of time when original file data in selected version was modified. Moreover, a feature PITR (Point-In-Time-Recovery), that utilizes the original file history, is available (figure 5). For selected moment of time T it is only necessary to find version file with the lowest version number N and with modification time after T. Data resuming process begins with making copy of original file. Afterwards, to obtain selected version N, all records from any snapshot file with version $M>N$ need to be read and processed in recursively executed compensation according to reverse version number order. Processing of single record assures that its data is written to original file at the offset determined from the content of the record. Single record checksum verification assures its data integrity.

Using original file history, an user is able to resume older versions of its data in two ways: by choosing version file or using PITR feature. All snapshot files are created automatically in the same directory where original file is stored. As a consequence, version and snapshot files could not be considered as a secure copy of original file.

5 Efficient Storage Use by Universal Snapshot File

A universal snapshot file concept, presented in this chapter, offers many functionalities like file data history or point in time recovery. This features use in recursion compensation process, in which many snapshot files could be processed. The recursion stop condition for the problem of restoring original file data to N version is defined by reverse order of compensation process using each time snapshot files with version number $M>N$. In single compensation process many I/O operations are executed. Most of disk operations in compensation process are read operations, but making copy of original file requires also write operations. This is a consequence of fact that universal snapshot file includes only records created earlier (according to the COW rule) and it is not a full data copy from original file.

Snapshot techniques are characterized by efficient use of storage space, because required snapshot data size is always not greater then original object size. Representing snapshot at file level is not as efficient as in case of filesystem or block-device level. Universal snapshot file concept utilizes filesystem resources (i.e. blocks or i-nodes) to store journal, snapshot and version files. Journal and snapshot files obviously have to store records; only the version file does not demand more and more resources (free blocks) regardless of its size. However, it has to emphasize that the excessive use of storage space is the cost of portability of universal snapshot file.

In universal file concept, a single record of snapshot file contains not only data, so (in the worst scenario) the size of this file may raise beyond the size of original file. In this disadvantageous case one or many records include all original data range. It is worth to mention that neither records from single journal nor snapshot file never have common original file data range, because each record holds its own separable data range. This is another important feature in terms of efficient storage space use. But this requirement implies also that writing record to journal or snapshot file is not always an incremental operation. Particularly, record merging may significantly increase duration of write operation belonging to logical transaction.

Presented universal file snapshot solution defines two main processes: synchronization and compensation. In synchronization process data are moved between journal file, snapshot file and original file. The proper order of operations is very significant, as it assures no data loss during whole process. All I/O operations of synchronization process might be executed in another order, discordant with request order. In order to avoid data loss it is necessary to use writing barriers that separate group of I/O operation. At first all operations, bounded to copy data from original file to snapshot file records, have to be executed. After finish all of those operations, it is possible to move journal file records to original file. It has to be noticed that the usage of writing barriers results in introducing additional latency in original file operations.

6 Summary

Universal file snapshot file technique, presented in this chapter provides a significant number of desirable features. Below we briefly describe the most important examples of these means:

- The presented solution is user friendly, because as all the activities are automatically taken by operating system. All necessary journal, snapshot and version files are created in synchronization process.
- Snapshot and version files are irrespective of filesystem, all snapshot data are stored as collection of records. Moreover, the way that version number is stored does not utilize any dedicated filesystem structures, so solution is portable between different filesystems. Provided that the original file is copied altogether with related snapshot and version files (where the relation is hold by version number), previous version of original data can be accessed in the new location as easy as they where accessible in the previous one. All original file data history may be copied or moved to other location, which in turn may be located in different filesystem.
- Every previous version of original file data is stored in snapshot file and is accessible through related version file. According to COW rule, snapshot file stores only the portion of original file data that was modified. When original file has its data history written in sequence of snapshot files, no pair of snapshot files stores the same records. Universal snapshot is not designed to serve as a backup of original file data, so storage space is efficiently utilized to provide access to previous versions of original file data.
- Any activity that is executed due to modifications of original file data can not cause data loss. The presented concept is based on logical transactions, that use journal file to assure that even in case when write error occurs, previous file

data could be accessible. Logical transaction protects original file data lost, even when logical transaction is rolled back due to concurrency problems with other transactions, its journal file is not removed.

- History of original file data modifications make it possible to restore any previous version, if only all needed snapshot files are available. Previous versions are accessed through version file using compensation process. Only directly previous version is available with single compensation. Otherwise compensation process is executed recursively with snapshot file read in reverse order determined by assigned version numbers. Reading original file data in version *N* requires an access to all snapshot files marked with version number *M*, where $N<M$. Is worth to mention that no dedicated tools are needed to read any previous version of original file data – this is one of the most significant differences from the way that common revision control systems operate.
- Version numbering makes it possible to determine creation sequence of snapshot files. This information is in turn required to keep original file data history. Point in time recovery allows user to determine a moment of time when original file data in version was modified. For selected moment of time one need only to find version file with the lowest version number *N* and with later modification time. Furthermore, PITR feature compensation process, executed recursively, uses single snapshot file.
- Assuming that logical transaction use auto-commit, which is defined by maximum fixed time limit between two following file modifications in the same transaction, external software is unaware of existence journal, snapshot and version files. Universal file snapshot solution is transparent to external software, although it is concerned with noticeably lower has performance level (compared to another snapshot concept) in case when process writes full data range every time file modification takes place.
- User is always able to remove version or snapshot file. As the result of this operation, all earlier versions of original file data will no more be accessible. The main reason of that phenomenon is the way that effective snapshot data are stored. Snapshot file data are relative to previous version of original file, which is accessible only if all the subsequent snapshot and version files are in place. Both journal and snapshot file contain records with data. In case that there is no free space in filesystem, no records can be appended to mentioned files. Therefore space reservation is demanded. Many operating systems make filesystem resource reservation for administrator and that spare space might be used if there is no free space available for the ordinary system user. Alternatively, a fixed filesystem space might be assigned by the operating system to create journal or snapshot file, but in compensation process this location is indispensable.
- The proposed concept of universal snapshot file does not provide any facilities to perform a snapshot of directories or group of files. It is able to operate only with a single original file. Directory contents depends on numerous files operations, in example it is modified by file remove operation It is noticeable disadvantage in comparison with other snapshot techniques.

It is noteworthy that the majority of the properties-mentioned above are advantages. Especially, that universal snapshot file prevents a file from unintentional data lost. Additional storage space usage is a cost of applying this solution, however having in mind modern block device capacities it should not be concerned as the important issue.

7 Future Research

Future research works are concerned to implementation aspects in example modern operating system, i.e. Linux. After the implementation phase is finished, universal snapshot file solution will be compared with other snapshot techniques working at filesystem or block-device layer. Comparison tests should involve at least the measurement of effective storage space usage as well as performance test (in terms of requested I/O operations).

Transactional file modifications and usage of COW rule to write records in snapshot files causes additional latency in file system operations. This is another aspect of future research, because latency level tends to be significantly various in different filesystems. Prepared model of file modification should include many scenarios, among others: parallel logical transactions processing, record merging in synchronization process and restoring file data in fixed version based on its history or using PITR feature.

References

[1] Watanabe, S.: Solaris 10 ZFS Essentials. Prentice Hall (2009)
[2] Aurora, V. A short history of btrfs. LWN (2009), `http://lwn.net/Articles/342892/` (accessed November 29, 2010)
[3] Smoliński, M.: System zarządzania przestrzenią dyskową – Linux Logical Volume Manager.In: XI Konferencja Sieci i Systemy Informatyczne, Łódź (2003)
[4] Weikum, G., Vossen, G.: Transactional information systems. Academic Press (2002)
[5] Garimella, N.: Understanding and exploiting snapshot technology for data protection, Part 1: Snapshot technology overview (2006), `http://www.ibm.com/developerworks/tivoli/library/t-snaptsm1` (accessed November 29, 2010)

It is noteworthy that the majority of the properties mentioned above are advantages. Especially, that universal snapshot file prevents file from unintentional data loss. Additional storage space usage is a cost of applying this solution, however having in mind modern block device capacities it should not be concerned as the important issue.

7 Future Research

[illegible]

Author Index

MIX
Papier aus verantwortungsvollen Quellen
Paper from responsible sources
FSC® C105338

If you have any concerns about our products,
you can contact us on
ProductSafety@springernature.com

In case Publisher is established outside the EU,
the EU authorized representative is:
Springer Nature Customer Service Center GmbH
Europaplatz 3, 69115 Heidelberg, Germany

Printed by Libri Plureos GmbH
in Hamburg, Germany